Relativism-Relativity

Relativism-Relativity: An Interdisciplinary Perspective on a Modern Concept

By

Maria-Ana Tupan and Marin Cilea

CAMBRIDGE SCHOLARS
PUBLISHING

Relativism-Relativity: An Interdisciplinary Perspective on a Modern Concept,
by Maria-Ana Tupan and Marin Cilea

This book first published 2013

Cambridge Scholars Publishing

12 Back Chapman Street, Newcastle upon Tyne, NE6 2XX, UK

British Library Cataloguing in Publication Data
A catalogue record for this book is available from the British Library

ISBN (10): 1-4438-4744-5, ISBN (13): 978-1-4438-4744-5

CONTENTS

LIST OF FIGURES

AUTHOR'S ARGUMENT

Setting out to write on "the many faces of relativism," Maria Baghramian admits that the subject can be carved up in different ways, its ascription depending "on a great variety of doctrines and positions" (Baghramian: 1), several of which have already entered into the history of ideas. At the same time, writers on the topic are reproached for a lack of consistency concerning their criteria for the classification of the objects and domains of relativism.

It is doubtful whether consensus about classificatory definitions will ever be reached, even when less slippery notions than "relativism" are being debated. Maybe a common divisor or an algorithm for the definition of the concept in its intensional meaning should take priority over its extension, as the list of "objects" and "contexts" will always be an open set. In Baghramian's classification below (Baghramian: 6) for instance, we could add religious relativism to the cognitive/ moral/ aesthetic taxonomy (δ is actually doubling up α) of objects of relativism, considering that even Saint Paul approved of the altar whose inscription read "To an unknown god" (Acts 17:23), which may be paraphrased as "To God Whoever He may be"….

α cognitive norms: truth, rationality, logic, epistemic standards (cognitive relativism, epistemic relativism)
β moral values (moral relativism)
γ aesthetic values (aesthetic relativism)
δ knowledge claims, worldviews, ontologies, systems of belief (cognitive, conceptual, and epistemic relativism)

We have opted, therefore, for an "essay," i.e. an informal approach, which traces the birth and modulations of a defining concept in the passage from medieval to modern history. Since the sixteenth century it has meant different things to different thinkers, its symptoms being complaints including cognitive uncertainty (Michel de Montaigne), epistemological fallibility (William Whewell), and ungrounded morality (Friedrich Nietzsche), amongst others. This, however, has been only one side of the coin, the other showing pluralism as the liberating premise for the mind's supervening to genuine knowledge, epistemic structures, and ethical attitudes. It is especially through relationism, Karl Mannheim's word for a special case in the declension of relativism, that modern epistemology has turned up more fruitful results in its search for representational regimes of being and meaning.

The concept of relativity is today automatically associated with Einstein's physics, but is born of the confrontation between Cartesians and Helmontians on the early modern battlefield of the philosophy of science. It was one of the favourite issues of essayistic speculation in the later nineteenth century when it came to be perceived as a more appropriate epistemic position than Kantian transcendentalism in a post-Nietzschean world, which had experienced the transvaluation of all values. Walter Horatio Pater saw in relativity the very spirit of modernity.

In its twentieth-century hypostasis, radicalized by existentialism and aggravated by the agnosticism of the Copenhagen School of Quantum Physics, relativism was both lauded, especially by vanguard artists and subversive social groups, and exposed as an amoral and dangerous attitude.

Our essay is a revisionary and historicist approach to the issue, which cuts across the disciplinary borders of science, philosophy, ethics and art. Earlier versions of current theories in the New Physics, such as relative time and space, the two (implicate and explicate) orders, the superposition of states, fractal geometry, or space-time, are traced back to the work of Kepler, Leibniz or D'Alembert. Such ground-breaking ideas, first launched in the tentative form of the essay, which acquired a quasi-canonical status in the French Encyclopaedia, referenced other disciplinary fields (psychology, ethics, social science), and entered into fertile negotiations with discursive and formal innovations in literature.

Present-day holistic approaches to the order of artefacts as semiotic manifold, inter-disciplinary and cross-cultural studies are rooted in the cultural anthropology of the Victorian Age, whose relativistic statements and attitudes invite second thoughts on its reputation for heavy handed moralising and strong inclination toward classifications and taxonomies. Thomas Young's double-slit experiment of 1801, which had demonstrated the wave nature of light in opposition to Newton's corpuscular theory, had also pressed home the idea that all knowledge is partial and provisional, Cambridge philosopher William Whewell urging scholars to seek confirmation of their theories in the history of science and across disciplinary borders. The "consilience of inductions from classes of facts altogether different" (Whewell: 230) did not suggest a way to group things together according to natural relationships but a new vista leading to **polyvalent logic.** Thomas Henry Huxley would follow it and make an even more explicit statement in this sense: "The rationale of contradictory opinions may with equal confidence be sought in history". (Huxley: 230). Even the apostle of the end-of-the-century aestheticism, Walter Horatio Pater, who could no longer abide by laws, logic, and archetypes in the biological and perceptual flux of differentiation, sought an anchor for the phantasmatic house of life in the "unity of the spirit" achieved through communication across disciplinary spheres which "breathe a common air" and "catch light and heat from each other's thoughts". (Pater: 11).

In a way, Hegel was right thinking that no further progress was possible in the philosophy of spirit after his phenomenology, which looks now like a monument to the tradition of transcendental idealism. All great generalizations about the absolute in relation to the real and the ideal, the issue of system and structure in philosophy understood as science were gradually replaced with an anxious awareness of the constructed nature of man's world and, implicitly, of its historicity. The germ of deconstruction planted by Hegel the moment he allowed the Absolute Spirit to descend to language as "its main element and habitation" (Hegel: 439) took roots and bloomed forth when language lost its Cratylic transparency in relation to its referent turning upon itself as a historical, conventional, arbitrary and indeterminate medium.

Pluralism and relativism affect all theoretical concerns, statistics replacing analytic self-evidence, logical consistency and rules for the operation of the intellect or even empirical evidence. Philosophy ceases to be "the project of the self-understanding of reason on its own terms, within its own structure as given by language" (Hendrix: 13). The philosophy of science turns to the semantic energies of all forms of social expression, negotiating meanings with all the discourses in the public sphere. As reason no longer controls perception, which, in light of empiricist psychology, is relative to the perceiver, it seeks confirmation from as wide a circle of social subjects as possible. The relativist thinker works with multiple and mutually-exclusive points of view (Leibniz: logical **and** existential definitions of identity), makes room for the previously excluded third, and gives up on fixed identities. "More is different" says Noble-Award winner Philip Anderson: interaction with the environment will change a system's Eigenstates (characteristic features), so that everything may be said to exist only in a field of forces and in perpetual transition to something else. The reversibility of implicate and explicate order unfixes the relationship between parts and whole in structural metaphysics. The real and the ideal are collapsed in superpositions of states, and the disappearance of the symbolic distance between Platonic *archē* and phenomena renders an aesthetics of representation impossible. As in Borges's parable "On **Exactitude** in **Science**", the map that reproduces the territory point for point is rotting away its superfluity. On the other hand, are there rules for correct setting in perspective or representation? Deleuze and Guattari finished off the demolishion of a myth originating in Plato that the intellect can reduce things to *paradeigma*, establishing an unfailing parallelism between the choric and the thetic, or work out "a logic of tracing and reproduction" (Deleuze and Guattari: 12). Differential calculus, the recognition of the difference that will never completely vanish between the landscape of facts and theoretic constructs of figures, marked the beginning of the disenchantment of the mind from its humanistic dream of taking the universe into posession. The prevailing studies in non-linearity suggest that contemporary physicists share the philosophers' opinion that "mimicry is a very bad concept, since it relies on binary logic to describe phenomena of an entirely different nature."(Deleuze & Guattari: 11). The map does not reproduce but adds up to the territory. The new relationism allows of connection despite heterogeneity: "any point of a rhizome can be connected to anything other, and must be. This is very different from the tree or root, which plots a point, fixes an order." (Deleuze &Guattari:7). Patterns keep forming in the absence of constraints from pre-existing structures. Essentialism yields to contextualism, being in itself to being for or with an otherness:

> We will never ask what a book means, as signified or signifier; we will not look for anything to understand in it. We will ask what it functions with, in connection with what other things it does or does not transmit intensities, in which other multiplicities its own are inserted and metamorphosed, and with what bodies without organs it makes its own converge. A book exists only through the outside and on the outside." (Deleuze & Guattari: 4).

Relations work changes in the related elements: a "capture of code, surplus value of code, an increase in valence, a veritable becoming, a becoming-wasp of the orchid and a becoming-orchid of the wasp." (Deleuze & Guattari: 10)

It is this integrated picture of the world of experience that we carry with us in our progress through life. Cognitive psychologists have identified four types of relations establishing among elements as they are recorded in memory: word-concept (word dog refers to the concept "dog"), concept-concept (dog-animal), concept-percept (word-image), word-word (collocating words, words forming semantic fields: dog-tail-bone-cat…) (Griffiths: 213). Experiments have proved that subjects remember best elements associated with contexts and words related through topic. Where lists of semantic features prove difficult to memorise, topic models activate the richest patterns of node activities, because the objects of our experience are inserted within latent mental structures which are activated by any element of the set.

The last type of relations, word-word, is the structural principle of semiotic aesthetics: a troping of the topological, an open-ended, lateral and rhizomatic linkage of signifieds sliding under signifiers. Man's world, that Heideggerian worlding of the physical universe, is a web of related meanings, an integrated model of culture in which no element exists in isolation from the rest. Integrated teaching, interdisciplinary teaching by topics, is the present institutionalization of that incipient semiotic understanding of being which Leibniz figured out as a holistic order of interlocked monads.

Relativist thought is a constant in modernity which otherwise meant accelerated change, uprooting of distinctions and opening of the floodgates to the liberating manifestation of differences, concomitantly with the proliferation of codes built through them in an increasingly complex civilization.

From *onta,* the reflector's gaze turned to *relata*. Whewell coined the word "palaetiology" for his art of correspondences: the past retrieved from present evidence, examination of artefacts "from nation to nation and from age to age", and collection of evidence from the treatment in common of sciences so that they "may reflect light upon each other." (Whewell: I/ 640). With Leibniz in the previous century, the mind had turned away from the stupendous machine of the universe towards the successive orders of its own making, articulated around "principles, maxims, and rules of procedure" (Ibid.). These protocols governing the articulation of discourses, which Michel Foucault would call "episteme" or regime of knowledge and discourse, are acknowledged as multiple and provisional, yet fatally normative, since at no time can the human mind jump over its own shadow, disciplinary validity in between scientific revolutions being assessed in relation to them.

Nowadays relativism has ceased to be merely a range of attitudes, from the positively assessed liberalism to the negatively assessed suspended judgement: "Everything is allowed", to quote Ivan Karamazov. In the hypostasis of relationism, it is the model we use to describe the emergent universe, globalization, or the workings of the mind, the way we read and write about books, the way we teach…

The essay looks at the concepts of relativity or relativism as they have fed into simulation models of being or of world building, books at either end of this time span, in East or West Europe, being read, not as fruit of personal genius or an individual's empirical psychology, but as knots in international networks of cultural semiosis and institutional exchanges.

References

Baghramian, Maria. 2004. *Relativism*. Abingdon & New York: Routledge.

Deleuze, Gilles & Felix Guattari. 1987 [1980]. *A Thousand Plateaus: Capitalism and Schizophrenia*. Translation and Foreword by Brian Massumi. Minneapolis, London: University of Minnesota Press

Griffiths, Thomas L., Mark Steyvers, Joshua B. Tenebaum. 2007. *Topics in Semantic Representation. Psychological Review.* Vol. 114 No 2, 211-244.

Hegel, G.W.F. 1998. [1807]. *Phenomenology of Spirit.* Trans. by A. V. Miller. Delhi: Shri Jainendra Press.

Heisenberg, Werner. 1956. "The Uncertainty Principle". In James R. Newman (ed.). *The World of Mathematics,* Vol. 2. London: George Allen & Unwin.

Hendrix, John Shannon. 2005. *Aesthetics & the philosophy of spirit : from Plotinus to Schelling and Hegel.* New York: Peter Lang .

Huxley, Thomas Henry. 1880 [1997]. "Science and Culture". In Alan P. Barr (Ed.). *The Major Prose of Thomas Henry Huxley*. Athens, Georgia: University of Georgia Press.

Whewell, William. 1847. *The Philosophy of the Inductive Sciences.* Second Edition. London:" John W. Parker. Volume I.

CHAPTER I

THE TIME AND SPACE OF MODERNITY

I.1. Newton and absolute time and space

Heisenberg's "uncertainty principle" may be considered the Copernican Turn in physics that marked the shift away from the classical world through the erasure of the dividing line between the subjective and objective aspects of the world. The physical description of phenomena in space and time was declared impossible, any measurement being relative to the observer's apparatus: "it is not possible to decide, other than arbitrarily, what objects are to be considered as part of the observed system and what as part of the observer's apparatus" (Heisenberg 1956: 1054). Even the time-honoured debate around absolute or relative time, showcased in the museum of the "classical world," centred on the physical concepts of space, time and motion. In the attempt to put his finger on the pivotal point of the quantum turn, Heisenberg was sharing the stereotyped representation of modernity as a well-proportioned and harmonious edifice, built on the firm ground of physical laws and determinism. At a closer look, however, the edifice reveals its tensed structure, polarized between Newton's absolutism and Leibniz's relativism.

Newton's first law of motion states that: "Every body will remain at rest or in a uniform state of motion unless acted upon by a force," which constrained him to define speed as the distance travelled per unit of time. But which time? The time pointed at by some clock, or absolute time? Did the first law of dynamics establish a certain idea of time? Unlike his contemporaries, Descartes and Leibniz among others—who considered time to be the order of successive events—Newton opted, in his laws, for absolute time, i.e. that which is mathematical and independent from whatever there is or happens in the universe. The space in which bodies move is also an absolute one, a passive container, distinct from the things within it. It goes without saying that, in such space, motion too is absolute.

In the Scholium opening of *Principia mathematica*, Newton defines the concepts of time, space, position and motion, which he uses in his treatise. The definition of the first two concepts is controversial and remains to be resolved.

> I. Absolute, true, and mathematical time, of itself, and from its own nature, flows equably without relation to anything external, and by another name is called duration: relative, apparent, and common time, is some sensible and external (whether accurate or unequable) measure of duration by the means of motion, which is commonly used instead of true time; such as an hour, a day, a month, a year.
>
> II. Absolute space, in its own nature, without relation to anything external, remains always similar and immovable. Relative space is some movable dimension or measure of the absolute spaces; which our senses determine by its position to bodies; and which is commonly taken for immovable space; such is the dimension of a subterraneous, aerial, or celestial space, determined by its position in respect of the earth. Absolute space and relative space are the same in figure and magnitude; but they do not remain always numerically the same. For if the earth, for instance, moves, a space of our air, which relatively and in respect of the earth remains always the same, will at one time be one part of the absolute space into which the air passes; at another time it will be another part of the same, and so, absolutely understood, it will be continually changed. (Newton: 6-12)

It is interesting to note that, although Newton decided to investigate real phenomena—those which can be observed experimentally—he chose to work with absolute quantities, which cannot be measured, since they have "no relation to anything external." Absolute space is conceived of as independent of bodies, and would therefore also exist in their absence.

In opposition to Newton's absolute time, Julian Barbour discusses "the end of time," as there is no connection between successive states of the universe which might very well be represented as a collection of pictures. Barbour sees classic space-time as a block of crystal (Barbour: 1), through which time flows like a river. Newton's assumptions about absolute space and time have proven to be merely formal. It is only relative positions and motions that can be measured experimentally, since they alone are a physical reality.

In classical mechanics, spatial and temporal relationships among objects (events) are expressed by an inertial reference system of coordinates to which a clock has been attached. Newton's first law, the principle of inertia, applies to this system. In mechanics, the speed of interaction transmissions is taken to be arbitrary, so that all the clocks in the universe can be synchronized, i.e. it is possible to build a universal clock. As a result, the length of a certain event is the same no matter what reference frame is used. Simultaneity is absolute, i.e. two simultaneous events in a particular frame of reference remain the same in any other. Finally, events in Newtonian time space can be ordered in time.

Mathematical equations, in an ideal space-time, do not vary with the sense of time. Were we to change t with -t (reversible time), they would remain unchanged. Looking in either direction from the present moment, we can determine both past and future events.

I.2. Leibniz and relative time and space

The rival theory to that of absolute space and time was a form of relationism supported by Newton's contemporary, philosopher and mathematician Gottfried Wilhelm Leibniz. From Leibniz's perspective, space and time were just orders of coexisting or successive events (Fig. 1).

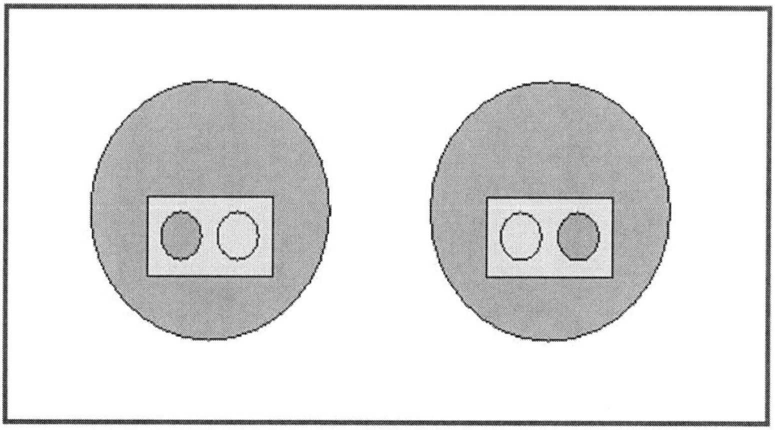

If space is real separately from substances, then two substances can be indiscernables and be in different locations – this entails a paradox. But, if space is a 'projection' of internal properties of substance, then two apparently indiscernible substances are really different by virtue of those properties which are perceived as spatial relations.

Fig. 1

Moreover, they are illusions rather than substances, depending on the existence of things in the world. Leibniz illustrates his concepts intuitively through an analogy with the members of a family in the fifth paper of his correspondence with Clarke (Leibniz 1716: 36-37):

> The mind can give itself a picture of genealogical 'lines', whose 'size' would consist purely in the number of generations, with each person having his place on one of the lines. Add to this the fiction of soul-migration, so that a single human soul could turn up a second time further down the line; if that were to happen, someone who was a father or a grandfather might become a son or a grandson, and so on. Yet those genealogical places, lines, and spaces, though they would express real truths, would only be ideal things. And here's another example to show how the mind is prompted by accidents that are in subjects to invent for itself something that is outside the subjects.

The family metaphor is classified as a "fiction" in this passage, rather than "generality," as Robin le Poidevin describes it in his *Travels in Four Dimensions* (38), by analogy with other abstract notions, such as "Papacy" or "Presidency of the United States": "not because there is some concrete item which persists through changes of holders of the office, but because different people at different times come to stand in the same relations to other people or institutions."

The distinction here—which took Bertrand Russell some time to resolve—is that between a class of objects and its elements. The notion designating the class, for instance, "family" or "felines," was eventually placed by Russell on a higher level of a hierarchical classificatory scheme than its members (tiger, panther, lion, etc). The distinction is logical, not ontological.

On the contrary, for Leibniz, space and time resemble here the innate ideas which are prompted by the accidents of Plato's reality of "shadows" (the objects of empirical experience). Immanuel Kant, who would define space and time as "representations," had his work cut out by Leibniz. While Kant separated the object of knowledge (epistemology) from the object of anthropology (representation, not according to reason, but "in a pragmatic or physiological sense," i.e. according to the individual's arbitrary associations), Leibniz separated the principle of essence (logical distinctions, observing the principle of non-contradiction) from the principle of existence. It is the latter that makes the whole of existence into a whole or a net of interconnected monads. As any predicate is said to be included in the subject, the infinity of virtual predicates stretches any individual entity to include the whole of existence. The source of predication about the entities existing in the world is observation (that which is discernible), as well as ideal representations for which there is no factual correlative; one such example is the fiction of incarnation, which unfixes types (class of objects) from origin, triggering the endless sliding of signifiers: from father to son, etc. Space and time are of the nature of perception:

> I don't agree that everything finite is movable. According to the hypothesis of my adversaries [Clarke and Newton], a part of space is something finite yet not movable. Anything movable must be capable of coming to be in a new state that is discernible from the one it began in; otherwise, the 'change of place' is merely a fiction. For the moving of a finite thing to make any change that can be observed, the thing must be a part of some other finite thing. (Leibniz 1716, 33)

There is no pre-existing container for the material universe, and no empty and fixed position. Being positioned is equivalent to being situated in relation to other bodies, including that of displacing or being displaced. It is the whole configuration that creates possibilities for placement within the observer's perceptual field, which is not, however, relative to the individual observer.

Place is not where God is but where God acts, Leibniz says in his fourth paper; by this he means that space is relational, but the network of relations is law-like and akin to an information structure. Its intelligible nature is suggested by the "geometry of situations" phrase coined by Leibniz to describe it. It is according to certain laws that the geometry of a given situation changes, which boils down to an inferential statement about the dependence of time on space (Leibniz 1716, 20):

> 41. Clarke contends that space doesn't depend on the situation of bodies. I reply that it's true that space doesn't depend on this or that particular spatial lay-out of bodies, but it is the order that makes it possible for bodies to be situated, and by which they have a lay-out among themselves when they exist together, just as
>
> **what Leibniz wrote:** *le temps est cet ordre par rapport à leur position successive.*
>
> **what that means:** time is that order with respect to their successive position.
>
> **what he may have been getting at:** time is the order that makes it possible for events to have a chronology among themselves when they occur at different times.

I.3. Relativity in the twentieth century

The end of the nineteenth century saw the rise of a new disciplinary field in physics: electromagnetism. Into it fed the contributions of physicians from the previous century and earlier; Maxwell, whom a number of equations were named after, synthesized their work. The question was now whether Maxwell's equations remain the same (invariant) for different frames of reference. In mechanics, the problem had already been solved; the equations of Newtonian mechanics remained invariant under a group of equations called "Galilean transformations." They were used when a mechanic event was studied in another frame of reference. On the contrary, Maxwell's equations were not Galilei invariant.

At that moment, the situation in physics was as follows:

- Absolute space and absolute time continued to be the basic concepts of mechanics.
- The distance between two points and the duration of a phenomenon were also considered to be absolute.
- The geometry of the time was Euclidean.

The hypothetical and elastic medium, called ether, was supposed to fill the planetary void, which supported the propagation of electromagnetic waves.

- The highest speed of interaction with transmission had been identified as the speed of light in the void.
- The speed of light had been proved to be independent of the inertial system. In addition, Henri Poincaré found out that Maxwell's equations were invariant under Hendrik Lorentz's transformations.

The world, however, was still known in its old frame, since neither the discovery of the electromagnetic field, nor that of Riemannian spaces had significantly destabilized it. It was Albert Einstein who, in 1905, introduced a new theory in physics known as the special theory of relativity. Entities lost their autonomy and their identity, depending now on a system of space-time coordinates which positioned them in a four-dimensional universe with time marked off as an imaginary number, while measurements became dependent on the reference system. Existence itself, the Kantian "Ding an sich," was a thing of the past. Absolute values were gone, making room for the differences resulting from relative evaluations.

The universal clock, which had promised to measure the time on Jupiter as accurately as on Earth, had been replaced with a clock which showed time according to its position in various locations on Earth. At the base of a mountain, the time of the space distorted by the gravitational field differed from that pointed out by a mountain climber's clock at the top.

It was a matter of decades before mankind was forced to radically revise the familiar representations of its portion of the universe as if it had moved to another system of reference. And yet, people were being told that the "novelty" was nothing other than classical physics. By whom? By the young representatives of the Copenhagen School who were claiming to have achieved the true Copernican revolution in physics.

Quantum mechanics was not just relativistic but purely agnostic. The results of experiments were not only relative to the observer but also inaccurate when several parameters were concomitantly measured. This strange theory is known as the "Uncertainty Principle," as it must indeed have been perceived at that time, for until then the scientific world had admitted its limitations without principles. According to Werner Heisenberg, it was impossible to measure concomitantly a particle's position and its momentum. Another difficulty was that of predicting the collapse of the wave function (the pointer state to which the multitude of a system's possible states would collapse).

The post-war era has been entirely dominated by the New Physics. The space-time of our galaxy has ceased to be the only universe. Not only was the ether proved to be a fiction but also the mystery of the correlations of quantum systems from a distance, without contiguity, has not yet been solved. The multiverse theory is spinning its versions of parallel or interfering pasts and futures with humans crossing them as copies.

"Inquietudes" is the appropriate subtitle of a chapter in "La cosmologie comme « manière de faire un monde ». Physique, relativisme et Irealisme," contributed by Aurélien Barrau to *Forme et Origine de l'Univers*, a book he edited together with D. Parrochia in 2010. Here is a passage from his radical criticism of the relativistic attitude taken to the verge of ethical and conceptual chaos:

> Quels rapports cette cosmo-mytho-logie, qui n'est qu'un condensa – ou un précipité – des ramifications multiples de la physique contemporaine, entretient-elle encore avec la vérité ? Non pas avec les vérités détournées et retournées, la fidélité à l'événement de Badiou ou le rapport à l'infini de Hirt par exemple, mais avec la vérité en tant que telle. Se pourrait-il qu'elle invite, avec une acuité sans précédent, à renoncer au concept même de vérité globalisante et totalisante? Contribuerait-elle à porter un coup de grâce à la vision « correspondantiste » qui, depuis Socrates et malgré les objections insistantes de ces derniers siècles (de Vico à Dewey en passant par Hegel, Marx, James ou Tarski), constitue toujours l'acception orthodoxe et rassurante du « vrai » L'époque ne se prête guère à ces

positions supposées dangereuses et presque unanimement décriées. A en juger par la violence avec laquelle toute velléité au relativisme (au sens le plus large, qui couvre le constructivisme, le perspectivisme, etc.) est balayée – non seulement dans le champ épistémique mais aussi dans les champs éthique et esthétique –, il faut à l'évidence conclure que quelque chose de grave se passe ici. Moins une colère qu'une angoisse: le relativisme fait peur à la manière de ce à quoi on ne souhaite pas faire face. Pas d'affrontement, juste un mépris. Un revers de la main, un haussement d'épaule, une sage distance à instaurer entre ceux qui savent et le magma confus et subversif des « relativistes ».

References

Barbour, Julian. 1999. *The End of Time. A Talk with Julian Barbour,* http://www.edge.org/3rd_culture/barbour/barbour_p1.html.

Barrau, Aurélien. 2010. "La cosmologie comme « manière de faire un monde .» Physique, relativisme et Irealisme," http://lpsc.in2p3.fr/barrau/aurelien/colloque_barrau.pdfgards philosophiques sur la cosmologie.

Heisenberg, Werner. 1956. "The Uncertainty Principle". In James R. Newman (ed.). *The World of Mathematics,* Vol. 2. London: George Allen & Unwin.

Leibniz, G. W. 1716. *Exchange of Papers between Leibniz and Clarke.* Leibniz 5: 8.viii, http://www.earlymoderntexts.com/pdf/leibclar.pdf

Newton, Isaac. 1934 [1689]. Newton's Scholium on Time, Space, Place and Motion, Scholium to the Definitions in the *Philosophiae Naturalis Principia Mathematica*, Bk 1. Trans. Andrew Motte (1729). Rev. Florian Cajori. Berkeley: University of California Press.

Poidevin, Robin le. 2003. *Travels in Four Dimensions.* Oxford: Oxford University Press, http://www.scribd.com/doc/48116186/Travels-in-Four-Dimensions

CHAPTER II

IDENTITY BETWEEN ESSENTIALISM AND CONSTRUCTIVISM

II.1. Physics and identity

There are no two identical things, Leibniz says, grounding his belief in the consistency and wisdom of the Creator:

> Nature does not contain any pair of real things that are indiscernible from one other; because if it did, God and nature would be acting without reason in treating one differently from the other; so God doesn't ever produce two pieces of matter that are perfectly equal and alike. (Leibniz 1716, 31)

In the classical world, there are indeed differences among objects of the same kind, such as, for instance, a set of chairs. We say that these objects, which differ from one another, are "discernible."

By contrast, the objects of quantum mechanics (elementary particles) are considered identical if they belong to the same type, such as electrons, neutrons, etc. Being identical, they cannot be distinguished among themselves. We say, with a term borrowed from Leibniz, that they are indiscernible. The problem of identity in quantum mechanics is a complex one, since the behaviours of quantum particles differ from those observed in the classical world, and these can no longer be ignored, as Werner Heisenberg remarks:

> Many of the abstractions that are characteristic of modern theoretical physics are to be found discussed in the philosophy of the past centuries. At that time, these abstractions could be disregarded as mere mental exercises by those scientists whose only concern was with reality, but today we are compelled by the refinements of experimental art to consider them seriously. (Heisenberg: 65)

This is an excerpt from a lecture given by Heisenberg at Chicago University shortly after the beginning of experimentation in quantum mechanics, in which he explored the concepts of identity, individuality, and discernibility.

Absolute identity is a relationship established between a thing and itself. X is the same as Y (are the same), if and only if whatever is predicated upon X is also true about Y. The Principle of the Identity of Indiscernibles goes back to Leibniz (Leibniz 1716):

> 31. Our reasonings are grounded upon two great principles, that of contradiction, in virtue of which we judge false that which involves a contradiction, and true that which is opposed or contradictory to the false; (Theod. 44, 169.)
>
> 32. And that of sufficient reason, in virtue of which we hold that there can be no fact real or existing, no statement true, unless there be a sufficient reason, why it should be so and not otherwise, although these reasons usually cannot be known by us (Theod. 44, 196).
>
> 33. There are also two kinds of truths, those of reasoning, and those of fact. Truths of reasoning are necessary and their opposite is impossible: truths of fact are contingent and their opposite is possible. When a truth is necessary, its reason can be found by analysis, resolving it into more simple ideas and truths, until we come to those which are primary (Theod. 170, 174, 189, 280-282, 367. Abrege, Object. 3).

When we say that particle A and particle B are identical, what we mean is that there is just one particle with two different names. An object can have one or more features, but each of them is a part of its identity. The identity of an object (entity) is only one and there is no such thing as an entity without identity. Closer to the present day, W. V. Quine maintains both the ontologically grounded and the relative definitions of identity:

> We cannot know what something is without knowing how it is marked off from other things. Identity is thus a piece with ontology. (Quine: 55)

The individuality of an object, on the other hand, is defined by a number of characteristic properties. To put it differently, we can attach a set of properties to any object, through which it acquires individuality. The individualities of two objects are defined by the differences that exist between the two sets of properties. Even two seemingly identical things, such as two coins which have the same value, size and weight differ from each other at least on account of their distinct positions in space (it goes without saying that objects are considered impenetrable). It follows that the two individualized objects are discernible. Macroscopic things, therefore, are individualized and discernible.

Could we say the same thing about a number of electrons? Quantum literature will answer invariably that, since electrons (elementary particles in general), as quantum objects, cannot be identified, they behave as identical particles when they partake of an assemblage. (Castellani & Mittelstaedt 1587-1594).

The particles' lack of identity and, hence, their indiscernibility, was ascertained almost simultaneously with the birth of quantum mechanics. Erwin Schrödinger, one of its founders, went so far as to say that the elementary particles that lie at the basis of the physical world are indiscernible.

> Even if you observe a similar particle a very short time later at a spot very near to the first, and even if you have every reason to assume a causal *connection* between the first and the second observation, there is no true, unambiguous meaning in the assertion that it is the same particle you have observed in the two cases. The circumstances may be such that they render it highly convenient and desirable to express oneself so, but it is only an abbreviation of speech; for there are other cases where the 'sameness' becomes entirely meaningless; and there is no sharp boundary, no clear-cut distinction between (the two types of circumstances), there is a gradual transition over intermediate cases … I beg to emphasize this and I beg to believe it: It is not a question of our being able to ascertain the identity in some instances and not being able to do so in others. It is beyond doubt that the question of 'sameness', of identity, really and truly has no meaning. (Schrödinger 1951, 17, 18).

The emergence of quantum mechanics has raised new problems around the concepts of identity, individuality and indiscernibility. In fact, the question was whether Leibniz's principle of the identity of indiscernibles applies to the quantum world as well.

In order to find out in what way quantum particles behave, let us imagine two particles and analyze the possibilities of locating them in two boxes (or in two states). First of all, it must be said that elementary particles are indistinguishable no matter how they are considered, i.e. classic or quantal, in the sense that they possess the same intrinsic properties, such as electric charge, mass rest, etc. In classical physics, the particles can form four configurations (Fig. 2): (1), (2) and (3).

Configuration (3) will split into two variants: one particle to the left, and the other to the right, and then they change places. This distribution actually conforms to the well-known Bose-Einstein statistics.

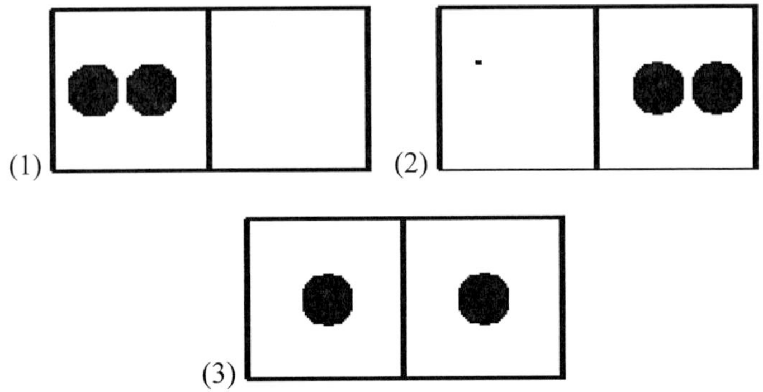

Fig. 2

According to quantum formalism, an ensemble of quantum particles has the property that the permutation of two particles leaves it unchanged. In other words, if at some moment the state of the ensemble is F, and, in this state, we exchange particles 5 and 14 among themselves, we get a state G which does not differ from F, in the sense that no measurement will identify some difference between F and G. We say that the system is permutation- invariant. Invariance under permutation is a property of quantum particles.

Things occur differently in classical physics. Any permutation will yield a different ensemble, so that, although particles are indistinguishable, they can be considered individualistic by virtue of some properties beyond the intrinsic ones. The fact that the permutation of quantum particles does not bring about changes in the state of the system leads us to the conclusion that these particles have no individuality and are non-individuals. H. Weyl found an intuitive and expressive way of describing this property:

> … the possibility that one of the identical twins Mike and Ike is in the quantum state E1 and the other in the quantum state E2 does not include two differentiable cases which are permuted on permuting Mike and Ike; it is impossible for either of these individuals to retain his identity so that one of them will always be able to say 'I'm Mike' and the other 'I'm Ike.' Even in principle one cannot demand an alibi of an electron! (Weyl: 241)

Bosons have the property that, in an ensemble, many of them may be in the same state. Bose-Einstein statistics does not prohibit that. That explains why two particles can occupy the same box (state), as shown in the Figure 3.

The fermions of an ensemble, however, are only distributed one per box. Fermi-Dirac statistics does not allow of two fermions in the same state. That is why their configuration is (3).

Apart from intrinsic properties (rest mass, charge etc.), classic and quantum particles are also possessed of state-dependent properties. Which properties are taken into account for the assessment of their indiscernibility?

The principle of the identity of indiscernibles takes three forms: (Krause & Coelho: 2).

a) The weakest form, P (1), which states that it is impossible for two individuals to share all properties and relations.
b) The stronger form, P (2) leaves out spatiotemporal properties.
c) The strongest form, P (3), only takes into account only monadic, non-relational properties.

One can notice that P (1) is not violated by classical particles, as classical statistical mechanics assumes such particles to be impenetrable, in the sense that their spatiotemporal trajectories do not overlap completely. We may conclude that they can be individualized by these properties.

Within the confines of classical physics, P (2) and P (3) get pretty mixed up. It is known that the particles of the same kind are regarded as indiscernible, i.e. they possesses all their intrinsic properties in common, and these are non-spatiotemporal. It is possible, therefore, for two individuals to share all their properties.

In quantum mechanics, things are different. If we believe that quantum particles are characterized by both intrinsic properties, and state-dependent properties, then even the weakest form of the principle, P (1), is violated. It cannot be used therefore to guarantee the individuality of a quantum particle falling back on state-dependent properties, as in the classic case.

John Stachel suggests an interesting analysis of the electrons present in a stable structure, such as the atom. Each electron is here characterized by a set of four quantum numbers, which is as much to say as that these particles will acquire some measure of individuality derived from their position inside a structure (atom). Identity, as *haecceity*, is not to be separated from its *quiddity*. Stachel defines as follows his notion of "structural realism," i.e. of a universe which is thoroughly structured:

> Before turning to some of the problems raised by the role of individuality in various physical structures, it seems proper to say what I

> mean by "structural realism." As discussed in Stachel 2002b, this term has been given various interpretations. For short (and with no claim of historical accuracy), I group these interpretations under three headings: Pythagorean, Platonic and Aristotelian. For the Pythagorean, mathematical structures are the only reality. For the Platonist, structures impose themselves on inherently (formless) matter. For the Aristotelian, matter is inherently structured.
>
> I would describe myself as a "genetic Aristotelian." That is, I take processes rather than things as fundamental. But these processes have a material basis. At any stage of its development, matter is inherently structured in many complex ways. This is the synchronic aspect of structure. But there is also a diachronic aspect. Any given structures are the result of preceding processes of structuration: they have a history of their formation, a limited duration, and their dissolution results in the formation of other structures. (Stachel: 207-208)

John Stachel's concept of identity is, therefore, no longer context-invariant. We have come to a relativistic position in the argument which makes every entity part of a "stupendous whole," whose birth is in the indistinguishable units—field quanta—of the quantum fields.

II. 2. Relative identity

Absolute identity means everything or nothing. One can speak of the identity of two objects only when the sets of their properties overlap; in fact, when the two objects are one and the same. There are, however, situations where one needs to describe objects which are partly identical, partially distinct, or relatively identical.

On the other hand, some philosophers criticize the notion of "absolute identity," because an elegant logical definition cannot be provided for it. Identity, therefore, is only relative, defined as follows: it is possible for two objects, A and B, to have the same P property, but not to share the G property as well. In the example, "X is the same chair as Y, but X is not the same material as Y," the two objects X and Y are the same in relation to a particular property (chair) but not with respect to the material they are made of as well. Such distinctions necessitate the operation of a sortal concept for the identification of an equivalence class.

Being more resilient, the notion of relative identity can offer simple solutions where absolute identity fails. It becomes apparent that, in the framework of this type of identity, one can only make relative statements. **An object ceases to be a single, precise and absolute identity**, defined on the basis of context-invariant properties. Instead of being tautologies of the selves, objects allow the observer to associate them on the basis of a limited number of common contextual features. An intelligible pattern replaces thus an atomistic, starburst picture.

We may even conceive of an object's identity as a set of relative identities belonging to a vector space, such as the Hilbert space used in quantum mechanics. In other words, the identity of the object is the totality of possible states generated by its relative identities. The disclosure of some particular identity in an object when set in relation to a context may be regarded as a process of de-coherence, when, of all possible states, only one pointer state will be realized at that moment.

To have a representation of the realization of relative identity states in a given space, we can imagine our three-dimensional space filled with identical cubes stacked next to one other in the three directions. Each cube contains a possible state of an object's identity. Together, these independently existing states define, through superposition, the identity of the object.

II. 3. Suspended identity: the self multiplied as me-copies across the multiverse

The Garden of Forking Paths by Jorge Luis Borges may be seen as the proleptic scenario of an endlessly forking universe, hatched in the mental laboratory of an artistic genius, but achieving the status of scientific hypothesis after the war:

> All these, however, are just metaphorical examples meant to color the grayness of technical explanations. A notable exception is "The Garden of the Forking Paths," where Borges proposes, unknowingly (there is no way he could have known), a solution to a still unsolved problem in quantum mechanics. "The Garden . . ." published in 1941, anticipates literally Hugh Everett III's doctoral thesis, published in 1957 under the title "Relative State Formulation of Quantum Mechanics" [Everett: 454] and that Bryce DeWitt later popularized as the Many Worlds Interpretation of Quantum Mechanics. (Rojo: 70)

At the close of the twentieth century, Everett's theory was boosted by mathematician Max Tegmark, according to whom any mathematical structure can create a universe.

The Fabric of Reality (the title of a book published by David Deutsch in 1997) is even more complex than that, with the author describing a multiverse which, unlike Everett's, forks both in space and towards innumerable futures. The alternatives at some decision point will grow, as if out of a severed sapling, into corresponding (consistent) pasts and futures.

Borges did indeed think of his fictional space as an unusual one, as his own invention in the "republic of letters," whose originality consisted in its forking out in time instead of space. There is, therefore, the possibility of the past or the future to be altered according to the protagonist's choices of alternatives.

A Chinese astrologist, Ts'ui Pên, gives up his official position as governor in order to write a book and build a very uncommon labyrinth.

Ts'ui Pên's book is a superposition of states, being itself and also the world which is written into being: a temporal labyrinth forking to infinity. Its chaotic nature comes from the amount of uncertainty in a universe which is a mix of determinism and randomness. People are free to choose, but, irrespective of their choice, both alternatives are realized in parallel words. Modal logic, which demands that one world exist to the exclusion of all others, does not apply here. In Borges' story, as in Everett's relative-states theory, one does not opt for one possibility but for all at once. Each alternative will extend into its own future and past.

English Sinologist Stephen Albert explains to Yu Tsun the maze structure designed by his ancestor, Ts'ui Pên, which he has managed to decipher:

> In all fictions, each time a man meets diverse alternatives, he chooses one and eliminates the others; in the work of the virtually inextricable Ts'ui Pên, the character chooses—simultaneously—all of them. He creates, thereby, 'several futures', several times, which themselves proliferate and fork. (Borges: 125)

A character wills a multiverse for the first time in literature: "I leave to several futures (not to all) my garden of forking paths." The limitation on the number of futures (several but not all), would suggest some laws of supervenience, and some constraints on the probability calculus which apparently concerns collocation:

> That fabric of times that approach one another, fork, are snipped off, or are simply unknown for centuries, contains all possibilities. In most of those times we do not exist; in some, you exist but I do not; in others I do and you do not; in others still, we both do. In this one, which the favouring hand of chance has dealt me, you have come to my home; in another, when you come through my garden, you find me dead; in another, I say these same words, but I am an error, a ghost. (Borges: 127)

The "Stream-of-consciousness" narrative structure is rooted in Bergson's idea of consciousness, which shuttles between the characters' *now* of experience and the associative *then* of memory. Borges was writing in advance of the physicists' realization that "in each quantum measurement the universe ramifies with one component for each possible result of the experiment." (Rojo: 75). Or, as Everett expressed it:

> The 'trajectory' of the memory configuration of an observer performing a sequence of measurements is thus not a linear sequence of memory configurations but a branching tree, with all possible outcomes existing simultaneously. (Everett: 454)

The amazing similarity between a piece of fiction and a physics thesis is explained by Rojo as negotiations among disciplinary discourses governed by a certain regime of knowledge:

> Borges' mind was immersed in the cultural matrix of the twentieth century, in that complex web whose secret components ramify beyond the demarcations between disciplines. (Rojo: 76)

In Rojo's reading of the famous story, Borges appears to have used J. W. Dunne's multiverse model with an infinite number of temporal dimensions and regressive observers in his *Experiment with Time* (1927), and with serial space in *The Serial Universe* (1934), whose impact on the literary imagination of the pre-war era deserves more attention than it has so far received. More traces of Dunne's multiverse can be identified in Flann O'Brien's own games with timelines, including the invention of parallel narrative trajectories, which have become a sort of mannerism in post-war novelists.

The ongoing war and the profession of the protagonist demand that his actions be coded. The murder of a famous English scholar is not motivated by the German agent's personal hate. His purpose lies completely beyond his victim's person. It is only Albert's name, the same as that of the city which is in the Germans' interest to bomb, that needs to pass on to them. According to his crooked logic, it was the news about Albert's assassination, appearing in newspapers in association with that of the murderer, i.e. their spy, which would convey to his boss the desired information. In other words, the spy has become a code operator, who sends coded messages, instead of reports in a natural and referential language. As in Tegmark's generational mathematical models, Yu Tsun's off-hand coding has a reality effect. "Albert" does indeed trigger the bombing of a city of the same name.

In the opening of the story, it is the choice of the day the bombing will take place on that determines the realization of one of the possible futures. Its postponement from July 24 to July 29 projects us into a post-forking, which no longer gives us access to the alternative from which the murder would have been missing. The spy, nevertheless, would have identified himself with the murderer, the slices of Yu Tsun copies being the continuing person. His identity is defined as the multiplicity of his copies in the temporal bolgia (pits) of the labyrinth, as are Albert and his surviving copy in the parallel future a continuous self.

Multiple identity as the totality of me-copies across the multiverse is woven by David Deutsch into his *Fabric of Reality* (1997). The Chinese ancestor's labyrinth resembles a Turing machine, i.e. a universal generator of virtual realities. Deutsch takes a step further away from the relativistic speculations around the speed-dependent encounters in Einstein's theory (see the "twins' paradox"). The possibility for a historian looking for Shakespeare to realize that he himself is Shakespeare (Deutsch: 258-9) wreaks havoc on the classical notion of identity. In this light, the self is seen as a multiplicity of copies spread out transversely, across parallel worlds, as well as longitudinally, in different life histories, each with its past and future.

As long as the historian imagined by Deutsch maintains his unitary self, he may be said to evolve within the frame of Einstein's relativity theory. By contrast, the superposition of states (the historian alter and the Shakespeare alter) transfers him to a quantum world, where his multiple identities have nothing more in common with Leibniz's indiscernibility.

> Copies need not be strictly identical in the sense of the **identity of indiscernibles** relativized to universes: All of my copies see a coin spinning in a coin toss, but an instant later half my copies see 'heads' come up, the other half see 'tails'. A distinction between copies, versions, and variants is at work here. Variants of me need not see the spinning coin. Versions of me see it though some of them see 'heads' and some 'tails'. The multiple identical copies of me all see the spinning coin. (Deutsch: 256)

After the first forking—the postponement of the attack—the spy gets to Ashgrove train station, where he is met by children emerging from a "shadow zone." To his amazement, they know who he is and that he is going to visit Dr Stephen Albert, which means that they are coming from his future. The children are giving him directions, telling him to turn left at each forking. We realize that the configuration is that of a circle, with the traveller as the final destination. As Lacan says in his paper read at the Seminar on *The Purloined Letter* (Lacan: 56), in the semiological order, the language user receives from the Other his own message in reverse form (interpreted). The Germans are capable to use the coded information conveyed by Yu Tsun, interpreting it as he had meant them to do. The spy has made himself understood in a code shared with his employers, but not also with the victim—at least so he thinks. For, although less apparently, Albert can see through him as ingenuously as if he had broken the code of the labyrinth-book.

It is the object of most interest to them, the Sinologist says, that language users will not mention, deferring it through a trope (signifier for another signifier).The ancestor had never mentioned the word "time," which was actually his obsession. Being mortal, he was denied infinite time, which he could only appropriate through the never-ending book and its probabilistic logic of relative states. It is prohibition that forces Tsun to substitute the Albert body for the Albert name (the very definition of the signifying act).

Albert mentions a story he had read in Pên's book: an army had fought and won the victory under very different circumstances. In both cases, it had been the strong desire to look heroic that had determined those warriors to fight fiercely and accept death serenely. They had acted according to values inscribed in society's symbolic order (the Lacanian Law of the Father) and internalized by them as an unconscious text. Tsun admits to himself that he did not like the Germans, but he had felt the desire to prove to those racists that an alien too could be clever and resourceful. The desire to be recognised by the other is stronger than love of life. One can only get an exalted idea about one's person from the other's gaze:

> Nowhere does it appear more clearly that man's desire finds its meaning in the desire of the other, not so much because the other holds the key to the object desired, as because the first object of desire is to be recognized by the other. (Lacan 58).

Tsun is not the only character alienated into a *locus copiae*. As Albert is listening to Chinese music, while Tsun is crossing *his* garden, we may suspect that his long exposure to another culture might have bred into him the desire to be admired for his wit, as well as serene acceptance of the denouement he could not avert.

II. 4. Identity/ Alterity

J. W. Dunne's *Experiment with Time* opens with a chain of associations meant to illustrate the "train of thought," of the same kind as that which allowed Leibniz to include the whole universe into the semiotic consciousness of a single individual. Unlike Leibniz, however, Dunne roots psychology in ontology. In light of recent theories about the four-dimensional universe, he places the "observer of all observed" in an extra dimension of the universe inhabited by the object of observation: "A self-conscious person is one who knows that he knows," he says, undertaking to represent his serial observer with the help of mathematics and geometry. (Dunne a: XII).

In addition, evidence is provided in support of the idea that dreams open gates to a distinct world, as they turn out to be prophesies that come true. There is no code at work here, as in Borges' story, but an anticipation of the wormholes which allow passage to other dimensions or universes. The law generating the serial observers is "mathematical regress," i.e. "the expression of some relation" (Dunne a: 151):

> 1.Every Time-travelling field of presentation is contained within a field one dimension larger, travelling in another dimension of Time, the larger field covering events which are 'past' and 'future', as well as 'present', to the smaller field.
>
> 2.The serialism of the fields of presentation involves the existence of a serial observer. In this respect, every time-travelling field is the field apparent to a similarly travelling and similarly dimensioned observer. Observation by any such observer is observation by all observers pertaining to the dimensionally larger fields.
>
> 3.The focus of attention in any field has the same number of dimensions as has that field, and is a dimensional centre of the foci of attention in all the higher fields.

In the revised 1934 version, the field of presentation for the higher observer, in the second form of the regress, is spatial and not time—the spatialized continuum of a tensed universe (Dunne b: 90).

> And, if you wished to describe three configurations as separated in space, you could say no more about them than we have said of Z, M and JV.*
>
> So, although we begin by using the entropy order of the three configurations to represent their time order, the result is the discovery that, in the second term of an inevitable regress, this entropy order represents order in an unsuspected dimension of space. And it is clear that whatever we may select, at the outset, to represent time order must represent, at the next stage, nothing but space order. In other words, time order must change to space order at each stage of the infinite regress of real time.

In *The Third Policeman* by Flann O'Brien, we see the narrator forming a series from representations which present some analogies on the basis of which a rule may be devised, so as to create a set reaching out to infinity. The inner homologies make them look like subsets in Hilbert spaces, or as self-similar regularities of the fractal kind. Von Koch's 1904 snowflake or Sierpinski's triangle are configurations in which parts are similar to the whole. O'Brien speculates both on fractals and nesting algebra (see the wooden chests of regressive sizes treasured by one of the three policemen), but the source of his "mock serialism" was probably Dunne.

The psychological observer with a four-dimensional field of attention —one more than the time traveller for whom events are past, present and future—differs from the one in the smaller field of presentation through the degree of awareness ("I know that I know"). As Dunne does not relate the various glances to embodied subjects or to the real, imaginary or symbolic orders, as Lacan does in his psychoanalytic reading of *The Purloined Letter* (Lacan: 10), identity is eventually blown up by his "experiment" in endless self-division and abysmal splitting of consciousness.

The regressive sizes of the policeman's wooden chests in *The Third Policeman* are mocking Dunne's theory with its inflated mathematical jargon, the policeman carrying the chests being actually on an inferior level of awareness, like the King or the Prefect of Police in Poe's story, whose glance, as Lacan says, sees nothing. He and his colleague are the victims of the third policeman's deceptive games. It is intersubjectivity and the symbolic order rather than the individual's autonomous, imaginary relationship to *Umwelt* that define a man's way of being in the world:

> For the travelling conscious observer GH is the only thing which, by its intersection with the reagent O'O", distinguishes in O'O" the place O wherein that reagent is capable of conscious observation. Omit GH, and there is no O. Similarly, when we pass to Fig. 10, we see that the travelling field 3, G'G'H'H' coinciding with conscious observer 3, is the only thing which, by its intersection with reagent 2, DFBE, distinguishes in DFBE a line GH wherein that reagent is capable of consciously observing, as at O. Omit G'G"H"H' from the diagram, and GH, containing O, vanishes. And so it goes on throughout the series, to infinity. In short, leave out the higher conscious and successive observer, and the lower observer ceases to exist as either conscious or successive, though there still remains an unnecessary and unjustified diagonal. (Dunne: 152).

Although an enthusiastic reader of relativity theory, Dunne is actually supporting a monadic theory of subjectivity. It was O'Brien who sensed the ontological crisis caused by the demise of the notion of "matter" and of hierarchical, centred, hegemonic structures, as well as the victory of constructivism over essentialism. The disputed theory of implicate and explicate order showed both to be interchangeable. O'Brien's characters rise or sink to other ontological layers; for instance, those in the student's book *(At Swim-Two-Birds)* "plot" against him, i.e. both conspiring and sending him to a lower level of consciousness as a character in his character's book. The narrator starts spinning versions of the plot instead of a tensed narrative (present, past and future):

> One beginning and one ending for a book was a thing I did not agree with. A good book may have three openings entirely dissimilar and inter-related only in the prescience of the author, or for that matter one hundred times as many endings. *Examples of three separate openings—the first:* The Pooka MacPhellimey, a member of the devil class, sat in his hut in the middle of a firwood meditating on the nature of the numerals and segregating in his mind the odd ones from the even. He was seated at his diptych or ancient two-leaved hinged writing-table with inner sides waxed. His rough long-nailed fingers toyed with a snuff-box of perfect rotundity and through a gap in his teeth he whistled a civil cavatina. He was a courtly man and received honour by reason of the generous treatment he gave his wife, one of the Corrigans of Carlow.
>
> *The second opening:* There was nothing unusual in the appearance of Mr. John Furriskey but actually he had one distinction that is rarely encountered—he was born at the age of twenty-five and entered the world with a memory but without a personal experience to account for it. His

> teeth were well-formed but stained by tobacco, with two molars filled and a cavity threatened in the left canine. His knowledge of physics was moderate and extended to Boyle's Law and the Parallelogram of Forces.
>
> *The third opening:* Finn Mac Cool was a legendary hero of old Ireland. Though not mentally robust, he was a man of superb physique and development. Each of his thighs was as thick as a horse's belly, narrowing to a calf as thick as the belly of a foal. Three fifties of fosterlings could engage with handball against the wideness of his backside, which was large enough to halt the march of men through a mountain-pass. (O'Brien: 1939).

O'Brien is to many critics one of the writers who took fiction from modernism to postmodernism, and this awareness of the distinction between the imaginary (perceptual centre of consciousness) and the symbolic (phenomenology of the nature and function of symbolic forms) testifies to the existence of such shifting concerns in the poetics/politics of the novel in the mid-twentieth century. The world is ontically evacuated —what world is this? Brian McHale queries in *Constructing Postmodernism* —while the fictional "construction" becomes intensely aware of the "complex layering of mediating forms and genres." (McHale 1992, 13).

Modernist poetics tended to give the Self priority over the Other, but a Self multiplied by rhetorical masks. Identity is stronger when confirmed in its truth by masks, as Oscar Wilde put it in his poetics of intentional subjectivity (*Intentions*). It is identity by proxy. The favourite troping is borrowed from Tarot symbolism. Set theory does apply to Aleph numbers, Aleph being the first card of the Tarot pack. Aleph is also the first letter of the Hebrew alphabet, which is a complex code of numbers, letters and images. Serialization matches substantial multiplicity and abstract numbers, the real and the ideal. The Tarot story itself is threaded on layers of meaning, forking in the end through the Neophyte's enlightening or his demise. A progress similar to the alchemical release of the spirit from matter may be seen in the Aleph-Yod-Qof trinity. Aleph is the archetype, the "peace that passes all understanding." It enters duration and existence through Yod, rising again to cosmic Qof. The Golden Dawn, the name of another esoteric society, misses Aleph's incarnations as Yod and Qof, being the union of generative and productive powers, the true self incarnated in the golden ring of personalities. In *The Waste Land,* they are the creators of artefacts, quotes from their immortal work being part of the great tradition which could still save Spengler's decaying West from moral and intellectual dissolution. Infinite creative energy, beyond all consciousness; thought, beyond knowing.

The gyres in the poetry of Yeats resemble the existential cone of light in which Aleph enters space-time as Yod, and the cosmic cone, beyond space-time. Everything and everybody is polarised between fall and redemption, whether the agenda of salvation concerns Ireland or the entire West. The existential self merges with a higher double: present and medieval Ireland, Ireland and Byzantium, Dante as historical person and poet etc, bourgeois, materialist culture, or the imaginative cycle of the moon. At the end of *The Waste Land*, the affined fire of the imagination has brought one more poet to the Purgatorio of art.

In *Finnegans Wake*, Anna Livia Plurabelle, living in real Dublin, is the first Mother Aleph (ALP), as incarnated Air (water). The Fire Water Mother Aleph is transfigured at the end as a rainbow, the light of the existential cone being decomposed into the colours of the spectrum. The Aleph-Lammed-Phay trinity symbolizes the unthinkable-life-death, running at the end back again to the Liffey, the river crossing Dublin. Death is actually the reconciliation of Time and timelessness in an archetypal, plural Belle who has left life/matter behind.

The mythic scenario yields to one of symbolic death, i.e. a rite of passage from body to sign *(sema*: Gr. for tomb). The onomatopoeic "mememormee" is the sound of Anna Livia's voyage as the letter "mem," the Hebrew letter meaning "water." We can decompose the word into Mem-Memor-Me: Hebrew *mem* for water, Lat. *memor* for "who remembers, who has a good memory," but also the name of a Roman usurper, a type which keeps moving throughout history and the novel as the "Humpheres-Cheops-Exarchas-Genghis Khan-Napoleon-Gestapowits-Chekars … Humphrey Chimpden Earwicker," protagonists of mankind's "Liber Lividus," and *me,* the accusative of the first person pronoun both in Latin and in English. Anna Livia remembers the site of trauma, her lips getting livid from the rapist's kiss, and frees herself from marital bond and bondage. The transfiguration scene reveals her as memory and symbolic body, who has left behind the "baser elements." The body of life—water, *aqua vitae*—is now the one that remembers and is conscious of its separate existence: "I, this watery body that remember me." Anna Livia Plurabelle is redeemed as Dunne's self-observer situated on a higher level and in a different dimension, as embodied memory doubling up and folding back upon itself as text: "river-run."

References

Borges, J. L. 1998. *Collected Fictions.* Trans. Andrew Hurley. New York: Viking.

Deutsch, David. 1997. *The Fabric of Reality*. London: Penguin Books.

Dunne, J. W. (a). 2001/ 1927. *An Experiment with Time*. Charlottesville: Hampton Roads Publishing Company, Inc.

—. (b) 1934. *The Serial Universe*. London: Faber and Faber Limited.

O'Brien, Flann. 1939. *At Swim-Two-Birds.* http://webcache.googleusercontent.com/search?q=cache:8WsiOXzZm3sJ:uncharted.org/frownland/books/flann_obrien/O'BRIEN_FLANN_-_At_Swim-Two-Birds)

Castellani, E. and Mittelstaedt, P. 2000. "Leibniz's Principle, Physics and the Language of Physics." *Foundations of Physics*. Vol. 30, No 10: 1587-1594.

Everett III, H. 1957. "Relative State Formulation of Quantum Mechanics." *Reviews of Modern Physics* Vol. 29: 454.

Heisenberg, Werner.1930. *The Physical Principles of the Quantum Theory.* New York, NY: Dover Publications, Inc.

Krause, Décio and Antonio M. N. Coelho. 2004. "Identity, Indiscernability and Philosophical Claims." http://philsci-archive.pitt.edu/1594/1/KraCoePC4.pdf

Lacan, Jacques. 2009/1978. "Seminar on 'The Purloined Letter" podpdf.com/ebook/lacan-seminar-pdf.html

—. 1977 [1959]. *Écrits*: A Selection. London: Tavistock.

Leibniz, G. W. (a) 1698. *The Monadology.* 31 - 33. Trans. Robert Latta. Corvallis: Oregon State University, http://philosophy.eserver.org/leibniz-monadology.txt

—. (b) 2007 [1716]. *Exchange of Papers between Leibniz and Clarke.* Leibniz 5: 8.viii. J. Bennett's website of Early Modern Texts, http://www.earlymoderntexts.com/pdf/leibclar.pdf

McHale, Brian. 1992. *Constructing Postmodernism*. New York: Routledge.

Rojo, Alberto G. 1985. "The Garden of the Forking Worlds. Borges and Quantum Mechanics." http://www2.oakland.edu/oujournal/files/9-garden.pdf

Quine, W. V. 1969. *Ontological Relativity and Other Essays.* New York: Columbia University Press.

Schrödinger, E. 1951. *Science and Humanism*. Cambridge: Cambridge University Press.

Stachel, John. 2005."Structural Realism and Contextual Individuality." In Hilary Putnam. *Contemporary Philosophy in Focus.* Ed. Yemima Ben-Menahem. Cambridge: Cambridge University Press.

Weyl, H. 1931 [1929]. *The Theory of Groups and Quantum Mechanics.* London: Methuen and Co. (2nd edition.).

Zurek, W. H. 2003. 'Decoherence, einselection, and the quantum origins of the classical.' *Reviews of Modern Physics* - REV MOD PHYS, vol. 75, no. 3, pp. 715-775.

CHAPTER III

RELATIVISM:
THE ORIGINS OF A STATE OF MIND

> "To regard all things and principles of things as inconstant modes or fashions has more and more become the tendency of modern thought."
> —Walter Horatio Pater

Pater's historicist reflection on time as a pageant of changing styles in the conclusion to his 1969 *Studies in the History of the Renaissance* came at the end of a journey back to the dawns of modernity, where he had witnessed, like a modern Mage, the birth of the spirit of relativism out of the encounter between alternative world pictures and narratives.

Although it was contemporary biology that had opened his eyes to the "fine gradations" of the generative plasma, it is doubtful whether the resilient house of life would have opened him the door to a fan of phenomenological pluralism. Newtonian physics or Wilhelm Wundt's physiological psychology, rooted in Immanuel Kant's anthropology, whose victory over the epistemology of the *Critiques* the Königsberg philosopher had probably neither wished nor suspected, were still charged with a good measure of determinism. Pater's aestheticism was seeking an alliance among the humanists of the Renaissance who had fought back the fundamentalist spirit which had lighted the pyres of the Inquisition, and were now hunting down their opponents on either side of the religious wars.

III.1. Essayistic argumentation: the generic frame of modernity's intellectual pluralism

Subversion of, or overt attack on authority is probably the most important item on the reforming agenda of the Renaissance. Actually, as in any passage to a new cultural phase, the narratives of the one preceding it are criticized in the name of an older authority. The pagan Greeks and Romans are appealed to in the humanists' attempt to temper the leftovers of the Middle Ages: religious fanaticism, scholastic dogmatism, or monarchic absolutism. What Michel de Montaigne achieved in this sense in the sixteenth century through the invention of a new type of discourse —the essay—was to reclassify, as philosophical squib or *philosophème* (Derrida, 363), the satire on knowledge undertaken by Erasmus, Barclay, Thomas More, or Christopher Marlowe. In the next two centuries, the deconstruction of knowledge traditionally cast in the format of treatise or esoteric lore prepared the stage for another kind of logic and type of argument. The essay was employed as a vehicle for work in progress, for the launching of new and provocative ideas, as well as the exploration of hypotheses, for interdisciplinary topics and even as the medium of the new *mathesis universalis.* The *en cyclos* lore of the Italian *prisca philosophia*, emulated, among others, by the Essex circle in London, was communicated within narrow groups of the initiated. Diderot's Encyclopaedia, with its essayistic items, made knowledge available within the new sphere of public discourse, open to critique and free appropriation. What had started as a provisional and unpretentious "attempt" was now crowned as an institutionalized discourse of the learned. It was through the essay form that those which were revolutionary ideas at the time, but commonly accepted in the last century, first made their entry into print. The essay also served as the catalyst of the generic mixture characteristic of modernity. In the sixteenth and eighteenth centuries, before the "dissociation of sensibility" and the Kantian separation into disciplinary spheres, the essay discharged its entire liberation potential. Afterwards it became specialized and institutionalized, its form falling under the constraints of classical rhetoric and argumentation.

The genome

Following centuries of hegemonic structuring of social and cultural orders, with well-defined and observed codes (implying constraints on selection and combination of elements), Montaigne's essays emerged out of an integrative effort bridging outward observation of the world and self-reflexivity, learning and sensibility, the public and the private, society's semiological order, and individual subjectivity.

Calling upon reason as "physic" for the madness that had caused three members of his family to be burned at the stake, Montaigne is trying, in *An Apology for Raymond Sebond*, to wake the mighty of the world to a sense of proportion; to press home a sense of relativism, philosophical/ methodical doubt or of the need to take thought before taking action.

> Is it possible to imagine anything so ridiculous as this miserable and wretched creature, which is not so much as master of himselfe, exposed and subject to offences of all things, and yet dareth call himselfe Master and Emperor of this Universe? (Montaigne: 225*)*

Montaigne was the defender of a treatise of Natural Theology going against "the authoritie of Lawes, or reverence of ancient custome." The target of social critique, however, is the wrong motive for conduct rather than the fallen condition of the estates. The following passage may have been the inspirational source of Hamlet's remark about the disproportion between motive and action in those who kill out of sheer vanity.

> How comes it to passe that so few are found who have still held one same wil and progresse in our publike revolutions, and that we see them now and then but faintly and sometimes as fast as they can headlong to runne into the action? And the same men, now by their violence and rashnesse, and now through their slowness demissnes, and heavines to spoile, and as it were overthrow our affaires, but that they are thrust into them by casual motives, and particular consideration, according to the diversities wherewith they are moved?

The hammers of critical judgment demolish the idols of social hierarchy and authority:

> Is it possible to imagine anything so ridiculous as this miserable and wretched creature, which is not so much as master of himselfe, exposed and subject to offences of all things, and yet dareth call himselfe Master and Emperor of this Universe? In whose power it is not to know the least part of it, much lesse to command the same. And the Privileges which he so fondly challengeth to be the onely absolute creature in this huge worlds frame, perfectly able to know the absolute beautie and several parts thereof, and that he is only of power to yeeld the great architect thereof due thanks for it, and to keepe account both of the receipts and layings out of the world. Who hath sealed him his patent?

The questioning of "privileges" and "patent" assumed by those demanding that everybody should obey and accept their faith unconditionally (*cuius regio, eius religio,* as John Donne is putting it in his third Satire) calls to mind the credential letters of ambassadors and of diplomacy as a new way of doing international politics. Truth too was to be negotiated in a society emerging as an architect's layout rather than as an uncontrolled sprawl.

The exercise of rational faculties, such as observation of plants and animals, had led to the astounding revelation, at least to the Aristotelians who had read *Historia Animalium,* that creatures inferior to men were capable of intentional, planned action. Accepted theories are turned on their head, and history is ransacked for persuasive evidence of the irrational cruelty of the mighty ones, an emperor who throws a slave to be devoured by a lion descending lower than the lion who spares his former healer.

> Why doth the Spider spin her artificiall web thicke in one place and thin in another? And now useth one, and then another knot, except she had an imaginary kinde of deliberation, fore-thought, and conclusion? We perceive by the greater part of their workes what excellency beasts have over us, and how weake our art and short our cunning is, if we goe about to imitate them.

The twilight of the earthly gods descends, bringing down emperors who are now level with the decerebrated spiders. The essay is corrosive, not so much of autonomous identities, as of divisions, categorical boundaries or ontological distinctions:

> The soules of Emperors and Coblers are all cast in one same mould. Considering the importance of Princes actions, and their weight, wee perswade ourselves they are brought forth by some weighty and important causes; wee are deceived: They are moved, stirred and removed in their motions by the same springs and wards that we are in ours. The same reason that makes us chide and braule and fall out with any of our neighbours, causeth a warre to follow betweene Princes; the same reason that makes us whip or beat a lackey maketh a Prince (if hee apprehend it) to spoyle and waste a whole Province. They have as easie a will as we, but they can doe much more.

The relativism of perception, the fact that actually both perception and apperception are relative to the perceiver or the phenomenologist/ interpreter, which would be the basic concern and ground of new disciplines from Immanuel Kant to Wilhelm Wundt, William James and Charles Sanders Peirce (anthropology, physiological psychology, pragmatism and semiology, respectively) are here stated for the first time in a nutshell.

> That things lodge not in us in their proper forme and essence, and make not their entrance into us of their owne power and authority, we see it most evidently. For if it were so, we would receive them all alike: wine should be such in a sicke mans mouth as in a healthy mans … Now if on our part we receive any thing without alteration, if mans holdfasts were capable

> and sufficiently powerfull by our proper meanes to seize on truth, those meanes being common to all; this truth would successively remove it selfe from one to another. And of so many things as are in the world, at least one should be found, that by an universall consent should be beleeved of all. But that no proposition is seene, which is not controversied and debated amongst us, or that may not be, declareth plainly that our judgment doth not absolutely and clearly seize on that which it seizeth: for my judgment cannot make my fellowes judgment to receive the same: which is a signe that I have seized upon it by some other meane then by a naturall power in me or other men.

The disunity of the self, its being in perpetual flux, mutant and passive, harbouring unpredictable psychological processes, is the final blow to any claim to transcendental forms of subjectivity. The object of thought becomes inseparable from the thinking mind, which produces maps of the world that never overlap. There is no *Mappa Mundi* to guide life's pilgrims on earth, only mapping as an endless process of the mind's folding of the world within itself.

> If I chance to take a booke in hand I shall in some passages perceive some excellent graces, and which ever wound me to the soule with delight; but let me lay it by and read him another time; let me turne and tosse him as I list, let me apply and manage him as I will, I shall finde it an unknowne and shapelesse masse. Even in my writings I shall not at all times finde the tracke or ayre of my first imaginations; I wot not my selfe what I would have said, and shall vexe and fret my selfe in correcting and giving a new sense to them, because I have peradventure forgotten or lost the former, which happily was better. I doe but come and goe; my judgement doth not alwaies goe forward, but is ever floting and wandering.

Writing will never be separated from the writing instance, as the qualities of things are in the mind of the perceiver, not in the object of perception, judging by the variety of impressions which different subjects get from the same things. Nobody has the right to impose on others his own image of the world:

> Whatsoever is knowne, is without peradventure knowne by the faculty of the knower: for, since the judgment commeth from the operation of him that judgeth, reason requireth that he performe and act this operation by his meanes and will, and not by others compulsion: as it would follow if wee knew things by the force, and according to the law of their essence. Now all knowledge is addressed unto us by the senses: they are our maisters … After all, we should know no more then a stone, unlesse we know that here is sound, smell, light, savor, measure, weight, softnesse, hardnesse, sharpnesse, colour, smoothnesse, breadth and depth. Behold

> here the platforme of all the frame and principles of the building of all our knowledge. And according to some, science is nothing else but what is knowne by the senses. Whosoever can force me to contradict my senses, hath me fast by the throate, and cannot make me recoyle one foote backward. The senses are the beginning and end of humane knowledge.

The distinction between the thing in itself, which is unknowable, and the perceiver as the source of all knowledge of the world, prefigures Kantian epistemology, but nothing in the way of a transcendental subject is called upon to sustain a higher order of signification. Montaigne confers this status on the symbolical order of artefacts. The aesthetic is placed above knowledge and faith:

> That the senses do often maister our discourse, and force it to receive impressions which he knoweth and judgeth to be false, it is daily seene. I leave the sense of feeling which hath his functions neerer more quicke and substantiall, and which by the effect of the griefe or paine it brings to the body doth so often confound and re-enverse all these goodly Stoicall resolutions, and enforceth to cry out of the bellyache him who hath with all resolution established in his mind this doctrine, that the cholike, as every other sicknesse or paine, is a thing indifferent, wanting power to abate any thing of soveraigne good or chiefe felicity, wherein the wise man is placed by his owne vertue: there is no heart so demisse, but the rattling sound of a drum or the clang of a trumpet will rowse and inflame; nor mind so harsh and sterne, but the sweetnesse and harmony of musicke will move and tickle; nor any soule so skittish and stubborne, that hath not a feeling of some reverence in considering the clowdy vastitie and gloomie canapies of our churches, the eye-pleasing diversitie of ornaments, and orderly order of our ceremonies, and hearing the devout and religious sound of our organs, the moderate, symphonicall, and heavenly harmonie of our voices:

Making its entry into the Renaissance ark of generic hybrids, suggestive of the artists' ambition to create a *genus universum* as the proper vehicle of universal learning, the essay opened like a fan outwards to the world of cognition and inwards to that of reflection, poising with grace on the vanishing line between philosophical statement and self-reflexive expression.

III.2. Renaissance esotericism

The relativism born of the mythosophic eclecticism of the Italian Renaissance had bred a tolerant attitude to different cultures, if not a deliberate cult of otherness as the touchstone of values and ideas. The European tour, the visiting aliens, the dialogue or divan on the topic of the differences among civilizations were all familiar scenarios spanning the period between the fifteenth and the eighteenth centuries. It is with a boost of relativism that we take note of the fact that Newton's or Locke's ideas, which were riding with their shocking novelty the high tide of mechanical empiricism, have fallen out of favour with contemporary epistemologists, while the "antiquated" Helmontians, alchemists, Rosicrucians or Cabalists recycling the heritage of the *prisci theologi* in the fifteenth to eighteenth centuries are being read for their surprising anticipations of scientific theories of relativity and quantum physics.

Between Erasmus, whose *De copia, Apophthegmatum opus* and *Adagia* got into the grammar school curriculum in England shortly after his lectures at Oxford University, and Count Volney, the ideologist of the French Revolution, a number of thinkers who were then less conspicuous than the rationalists and the luminaries lent the philosophical discourse the sound of tentative speculation, sceptical of the mind's ability to dispel the inexhaustible mystery of the universe through discovery of effective causes, as well as of totalizing myths, such as the universal character of physical laws, the universe as perfect machine, absolute time and space. Persuaded of the provisional and historical rather than universal and timeless truths of man's theories and systems, they chose to work from within pluralistic frames of reference.

The resurrection of Alexandrian eclecticism through Marsilio Ficino and Pico della Mirandola, the mix of Neo-Platonism, Jewish Cabala, Gnosis, Hermeticism and Alchemy created a space of free play for the mind, which Protestant reformers of various factions filled with monomyths reminiscent of medieval fanaticism. The scholastic jargon was sometimes employed by prudential subverters in order to beguile licensers, as Paracelsus may have done, but a middle-stream discourse is to be distinguished between the sermonizers who kindled the Puritanical spirit of Cromwell's New Model Army, on the one hand, and the dry language of empirical and close observation employed by experimenting Newton, on the other. As the latter is credited for the invention of scientific discourse, so did van Helmont create a stylistic matrix fit both for alchemy and chemistry, for mythology and science, which were still amalgamated in the discursive rites of passage to modernity.

Dialogism, as in Julia Kristeva's reading of Russian formalist Mikhail Bakhtin, is subversive of authority, which is forced into a decentred or even upside-down, carnival world, ridden with ambiguities and mistaken identities.

The Puritans' rejection of the apocrypha and return to "ancient worlds" and "biblical supremacy" (Bozeman: 26) had also meant a return to the letter of the Old Testament and the collapse of a whole library of medieval rhetoric on the doubly-articulated language of the Scriptures, on the distinction between *voces* and *subscriptio*, or between literal meaning and the three-level hierarchy of meanings which took the *sensus spiritualis* to the ultimate, anagogic interpretation.

On the contrary, the humanists' "return" to the pagan antiquity was a hermeneutic and philological effort to revive "bonae litterae," i.e. to remove the overlaying (what Heidegger would call "Destruktion") of scholastic interpretations in order to bring up the unadulterated original meaning. Ancient models were modified and added to present aspirations towards generic compounds or even *genus universum* and to the expansion of knowledge to an ark of universal lore (*mathesis universalis*). Analogy was weaving the meaning on different levels, sometimes the device being metalepsis (the troping of a previous trope) and other figures of transformation and substitution.

Leonardo da Vinci, for instance, embedded his representation of the divinely proportioned male body within a similar figure designed by Vitruvius, an ancient Roman architect who, in his turn, admitted to having proceeded by *analogia* with the three Greek orders: Doric, Ionic and Corinthian. Da Vinci's choice of an architect's creed was probably justified by the latter's educational model, which the Renaissance artist was also emulating:

> The architect should be equipped with knowledge of many branches of study and varied kinds of learning, for it is by his judgement that all work done by the other arts is put to test … (I/ I/1). Let him be educated, skilful with the pencil, instructed in geometry, know much history, have followed the philosophers with attention, understand music, have some knowledge of medicine, know the opinions of the jurists, and be acquainted with astronomy and the theory of the heavens.(Vitruvius: I/I/3)

Da Vinci modified the precedent, superimposing two male figures inscribed both in a circle and in a square. Unlike the original, centred on the naval, da Vinci's pose and geometry of perfect proportions privilege the head, the seat of Reason. The drawing actually displays sixteen different poses, which are as many views and ways of setting Man in perspective.

Vitruvius had also shown preference for the Corinthian order, which, unlike the Doric and Ionic, is rooted in a legend which ascribes it a hybrid and metamorphic nature, a cross of nature and artefact:

> A free-born maiden of Corinth, just of marriageable age, was attacked by an illness and passed away. After her burial, her nurse, collecting a few little things which used to give the girl pleasure while she was alive, put them in a basket, carried it to the tomb, and laid it on top thereof, covering it with a roof-tile so that the things might last longer in the open air. This basket happened to be placed just above the root of an acanthus. The acanthus root, pressed down meanwhile though it was by the weight, when springtime came round put forth leaves and stalks in the middle, and the stalks, growing up along the sides of the basket, and pressed out by the corners of the tile through the compulsion of its weight, were forced to bend into volutes at the outer edges. (IV/I/9).

We have been ushered into the modern world of AS IF, of huge analogies cutting across time, space and even ontological boundaries. The hermetic monad, Van Helmont's "archons," or Kepler's archetypes were all hypostases of the Spirit incorporated into the world of matter, which could be released through the alchemist's work, as well as gnostic asceticism, Neo-Platonic cult of Afrodita Urania, or Kabbalist return to the En Soph. Such concepts surrounded by a theosophic aura were mixed up with precise descriptions of movement or momentum or of chemical processes taking place in nature. As shown in Da Vinci's *Vitruvian Man*, the cosmos is built by the hermetic workman, according to analogies between what is below and that which is above. The Divine Proportion is no longer the work of God possessed of a lead rope (Amos 7:8), compasses (Prov. 8: 27, Job 27: 10), and square, as represented in Blake's 1794 watercolour etching entitled "The Ancient of Days Measuring Time" (Fig. 3), but the coextension of the whole space with the mind and body of a geometrician of the Pythagorean School.

What God seems to be measuring in Fig. 4 (*Bible Moralisée*, Codex Vindobonensis 2554) is the flower of life, also known as the "egg of life" or the "seed of life." Widely circulated throughout the Middle Ages, especially in Saint Omer's *Liber Floridus* (1120) —an encyclopaedic work based on a fifth century Latin original (Martianus Capella's *De Nuptiis Philologiae et Mercurii*)—the figure (Fig. 5) seems to represent both archaic Asian and European architecture, and the evolutionary dynamic of an original embryo, including the expansion of the universe out of an original core.

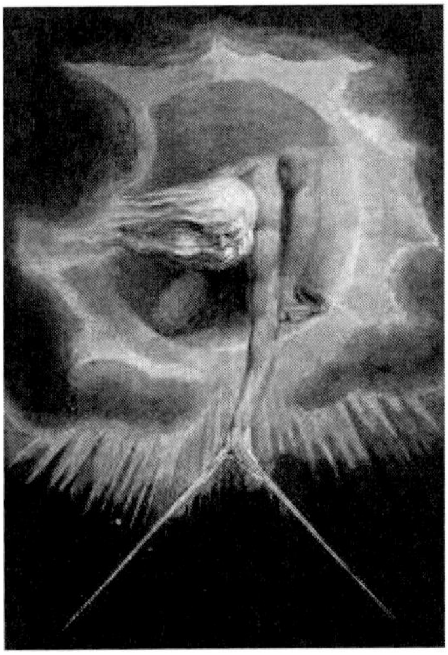

Fig. 3

Fig. 4

Fig. 5

The flower of life was studied by Leonardo da Vinci, being perhaps his inspirational source for a modified Vitruvian Man (Fig 6); it became a structural principle in Renaissance, the Baroque and neoclassic art.

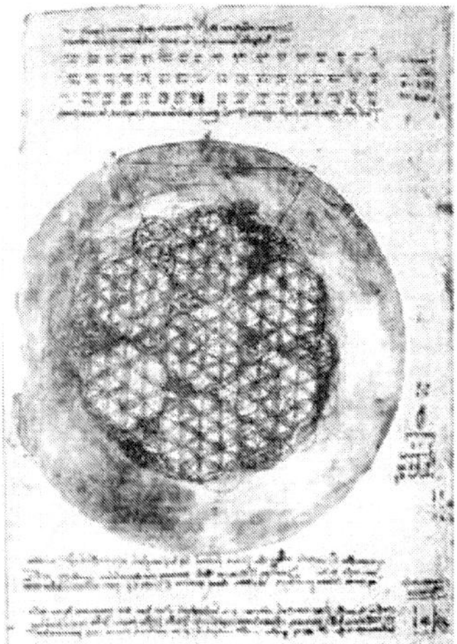

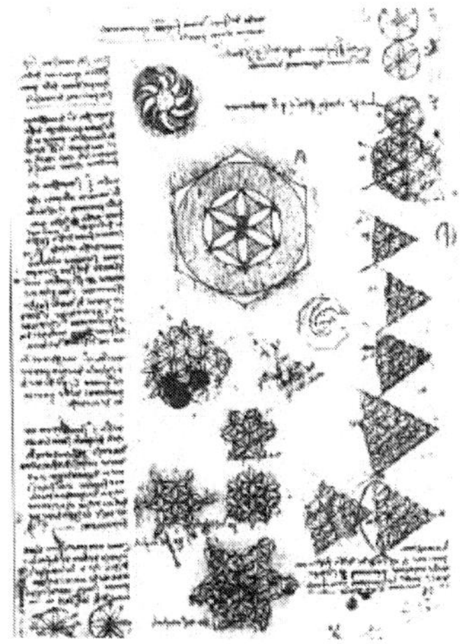

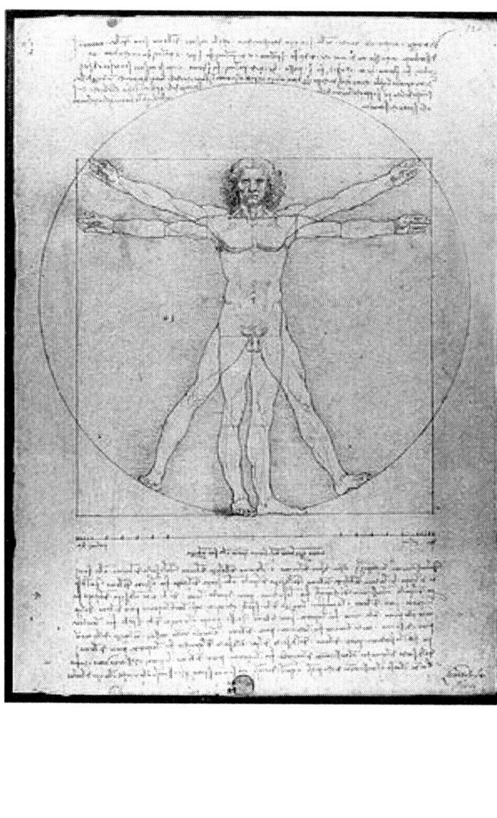

Fig. 6

The algebraic practice of ascribing numerical symbols to certain characters who were also carrying symbolic meanings on a physical, human and allegorical plane goes back to the first book of the Kabbalah, the Book of Creation (*Sefer Yetzirah),* composed, according to some sources, between the first and eighth century BC, and, according to others, between the third and sixth century AD. The Vitruvian Man too can be associated with the Adam Kadmon of the Zohar, a Kabbalist book of the 13th century, who was the first emanation of God, and in whose image was completed the whole of Creation.

Here is a "Daily Practice" which we can read in a manual of the School of the Holy Spirit (Schweighardt 1788):

> The outer and inner Mind
> Without God's light you cannot find
> Only the Spirit alone knows
> Reason in the flesh is blind.
>
> I. GOD
> God is free everywhere
> Within and without all Creature
> Time measure of Nature
>
> II. FATHER.
> God is the Alpha and Omega
> The Beginning and the End
> Time measure of the Law
> Lion with six wings
>
> III. SON.
> God is the first and the last.
> Time of the Evangelism
> Ox with six wings.
>
> IV. THE HOLY GHOST.
> And there is no God
> But the one God
> Time of fulfilment
> Eagle with six wings

The four figures dogmatized in the stanzas can be recognised as those from two Cabalist texts in the Bible: Ezekiel 1: 1, and Revelation 4: 1. A chimera, seated on the throne, reveals to the two prophets its four faces: a man, a lion, a bull, and a vulture. The man is here replaced with an Angel, whose etymology refers us to the Greek word for Messenger. He is the initiate in Hermeticism, who comes out of the darkness of the body, for, as long as his mind is its prisoner, bereft of "the light of God," it is impotent.

The Bull is the emblem of Christ, the "time of the Gospel," while the Eagle is the time *kairos* of fulfilment: the return to heavenly Jerusalem, the onset of the kingdom of the Spirit, the bird being the emblem of the Holy Ghost.

The circle and the Crescent make up the symbol both of alchemical bismuth and of the Bull's sign (Taurus) in the Zodiac. The symbolism is a complex one: the Moon and the Sun are reunited, the alchemical wedding preceding the revelation of the stone (*lapis philosophorum*). In medieval alchemy, lapis is Jesus. The Taurus is ruled by Venus, being, however, an Earth (Gea, Mother) sign of the Zodiac. Both symbolize Jesus—as Love and as the incarnation of the divine spirit on Earth out of love for man.

The Moon governs the Scorpio sign, and, in Alchemy, it is Silver: *argentum vitae,* the philosopher's stone.

The lion is a fire sign, ruled by the Sun father figure. In alchemy, it is represented by a circle with a dot inside. It is that point whose vibrations cause the universe to explode, and oscillate between expansion and contraction. Jesus is both the Father and the Holy Ghost, made incarnate on Earth.

The Renaissance man is defined as titanic aspiration to universality: to know everything, to travel everywhere, to win the race against time.

In the early Baroque era man's desire to break out of all boundaries can still be seen in various esoteric traditions, or in the cross fertilization of the arts, especially literary and painterly. It makes itself heard in musical flourishes as in the fugue's suggestion of perpetual motion. It deserves to be admired in the earliest disciplinary synthesis.

III.3. Kepler between Nothing and Everything: "On the Six-Cornered Snowflake," or Auguries of Fractal Geometry

With his informal essay, *On the Six-Cornered Snowflake*, Johannes Kepler secured a place for himself in the history of physics as the first explorer of crystal structure in search of the "formative principle" responsible for the generation of snowflakes. He intuited the underlying physical process which is the vapours' loss of heat, while his theory that mathematical structures lie at the basis of the universe has become current among physicists of the contemporary age, including astrophysicist Max Tegmark (Tegmark: 101-150).

Despite his remarkable intuition and scientific accuracy, Kepler chose an essayistic, playful and self-reflexive form. He relates the swift-vanishing snowflake to his precarious financial situation at the time he was working as the imperial astronomer for Emperor Rudolph II. The title is a pun on "nix" which, in his Low German, means both *snowflake* and *nothing*. The famous astronomer and mathematician saw the snowflake as an image of himself, an abyss of nothingness, unable, because the Emperor had forgotten about his pay, to give a friend a Christmas gift. Those "who have Nothing and receive Nothing have to give but Nothing?" Nevertheless, he thinks that the six-cornered snowflake would have made a perfect Christmas gift, "since it comes down from heaven and looks like a star."

Sticking to its legacy, of proceeding from sensuous observation to abstraction or intelligible representation, the essay reflects on the possibility of mapping real spaces onto geometrical figures, as well as on the nature of the agency which forces all snowflakes into the six-cornered shape. He sought out regular shapes which could fill the plane and their natural correspondences, such as the beehive or the dodecagon. (Fig. 7)

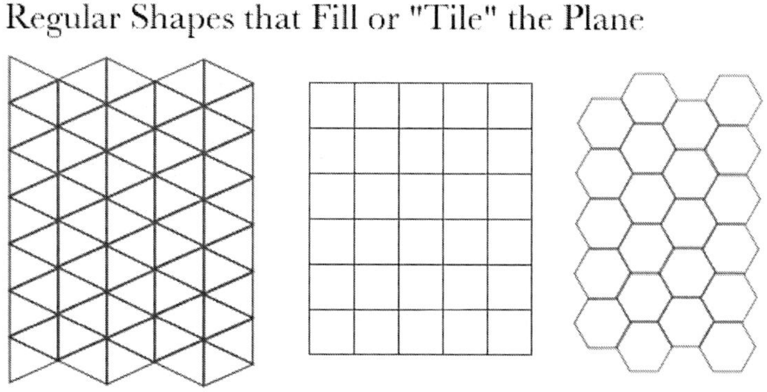

Fig. 7

Unlike Kepler's explorative attempt at self-similarity, the generation of the same figure was presented by Swedish mathematician Niels Fabian Helge von Koch in purely mathematical terms in 1904 (*Sur une courbe continue sans tangente, obtenue par une construction géométrique élémentaire*).The scientist's mind is working within the confines and under the constraints of one single discipline. Here is the purely geometrical "genesis" which no longer carries Kepler's heterogeneous and imaginative associations (Fig. 8).

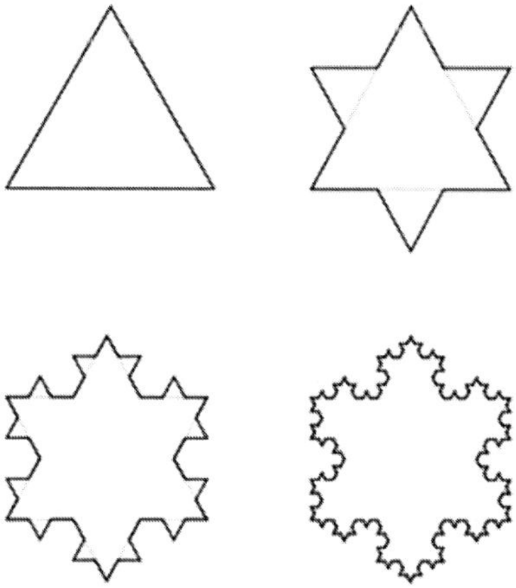

Fig. 8

III.4. Mapping chaos: Gottfried Wilhelm Leibniz, Laurence Sterne and differential calculus

The priority of Leibniz in the elaboration of differential calculus is not, maybe, the most important issue when it comes to the heated arguments he had with Newton, sometimes through an "included third" (Samuel Clarke). Whereas Newton, in *Principia mathematica*, limited himself to mathematical symbols, Leibniz produced a mix of mathematics, philosophy and psychology within an essayistic, speculative frame. Despite the apparently superior scientific prestige of Newton's treatise—he is actually credited with the invention of scientific discourse—it is the symbols employed by Leibniz that are used in differential calculus today; it is his philosophy of relative space and time instead of Newton's absolute space and time that was ratified by twentieth-century physics, and it was Leibniz again who proleptically proclaimed the existence of a monistic universe, in which each element contains the information of the whole. Fractal self-similarity, or the holographic plate whose whole image, in case the plate gets broken, reappears in each splinter were serious blows to the classical concept of identity as self-similarity and self-containment. Non-locality, demonstrated by the correlated changes in quantum systems that are separated in space, which is still perceived as a spooky form of relationism, seems to have imparted to that early modern philosopher a sense of serene connectedness to the All:

> 65. And the Author of nature has been able to employ this divine and infinitely wonderful power of art, because each portion of matter is not only infinitely divisible, as the ancients observed, but is also actually subdivided without end, each part into further parts, of which each has some motion of its own; otherwise it would be impossible for each portion of matter to express the whole universe.
>
> 66. Whence it appears that in the smallest particle of matter there is a world of creatures, living beings, animals, entelechies, souls.
>
> 67. Each portion of matter may be conceived as like a garden full of plants and like a pond full of fishes. But each branch of every plant, each member of every animal, each drop of its liquid parts is also some such garden or pond.(Leibniz b: 65-67).

Those who use the argument of non-locality nowadays in order to undermine the concept of identity would not probably invoke distribution as the only way of distinguishing particles among themselves. In Leibniz, we come upon both. He goes to all extent to impose a relativistic picture of being, probably in order to justify, as a royal ally, the *status quo* as the best of all possible worlds. Be it as it may, the way he anticipates present discussions on identity is absolutely remarkable.

Individuation, as his Philosopher glosses on it in *A Philosopher's Creed,* is dislodged from traditional arguments built around principles, causes, essence, substance ... Identity may be purelay relational:

> Here we touch upon the extremely thorny discussion of the **principle of individuation**, that is, the problem of telling the difference between identical objects, i.e. objects which differ in location alone. For example, two eggs could be so similar that not even an angel could distinguish them (assuming they were perfectly alike) but who can deny that they are different eggs? At least they differ in that one egg is *this egg here* and the other is *that egg there*. Thus we distinguish them by specifying one egg as "this one" and the other as "that one"; in other words, they are different because they can be assigned different numbers (indexed). Counting is just specifying one object after another; so there is no difference whether we count "One, two, three, ..." or whether we say "This one, and this one, and this one, ..." This specification is equivalent to specifying the time and place of the egg. We could specify it by its motion relative to some object (like ourselves) or by a motion of our own, like pointing at it. Therefore, the specification of identical objects depends on something external and not on the identical things themselves. (Leibniz c: web).

For Leibniz, knowledge of the world was permanently related to the knowing or, rather, perceiving mind. The differential calculus meant the filling of the physical landscape with decreasing geometrical shapes so as to approximate the former's irregularities as close as possible. The operation implied, however, psychological rather than purely mathematical factors:

With Leibniz, differential calculus is both mathematical and psychological, as the differentials of a curve are differentials of consciousness. Matter is infinitely divisible; it has no unity. The subjective I is the one that possesses unity, and acts as a multiplying factor, since the organic is finite, whereas the inorganic consciousness may achieve as many mappings of the former (inorganic inflation) as it pleases. The reasoning mind is the simple (unitary) substance:

> 17. Moreover, it must be confessed that perception and that which depends upon it are inexplicable on mechanical grounds, that is to say, by means of figures and motions. And supposing there were a machine, so constructed as to think, feel, and have perception, it might be conceived as increased in size, while keeping the same proportions, so that one might go into it as into a mill. That being so, we should, on examining its interior, find only parts which work one upon another, and never anything by which to explain a perception. Thus it is in a simple substance, and not in a compound or in a machine, that perception must be sought for. [...]
>
> 18. All simple substances or created Monads might be called Entelechies, for they have in them a certain perfection (echousi to enteles); they have a certain self-sufficiency (autarkeia) which makes them the sources of their internal activities and, so to speak, incorporeal automata. (Theod. 87.)
>
> 19. If we are to give the name of Soul to everything which has perceptions and desires [appetits] in the general sense which I have explained, then all simple substances or created Monads might be called souls; but as feeling [le sentiment] is something more than a bare perception, I think it right that the general name of Monads or Entelechies should suffice for simple substances which have perception only, and that the name of Souls should be given only to those in which perception is more distinct, and is accompanied by memory. .(Leibniz b: 17-19).

Under the guise of defending faith, Montaigne had actually dismissed religion from the sphere of rational examination or empirical evidence. The mythical frame, still preserved by Kepler in his layout of the Graz castle, by John Dee or even by Isaac Newton, with his odd penchant for alchemy, collapsed or was totally revised in the Age of Reason and of the Enlightenment. However, the emergence of an autonomous literary sphere in the seventeenth century consorted with a need, on the part of writers, to find some surrogate central *Sophia* as intellectual and moral centre of their modern fables. That "brought them within the sphere of influence of intellectual paradigms developed in non-literary fields of interpretation." (Moriarty: 6). Political as well, we might add. Sectarian ethic of work, labour on nature and self, Whig cult of the civilizing mission and of international flows of capital slot Robinson Crusoe in the penitent narrative of a redeemed citizen who has earned himself a place in a constitutional monarchy, after exchanging the discourse of monarchic absolutism with that of the colonial governor whose conduct is guided by the principles of instrumental rationality. The places he colonises catch the code of his own. On the contrary, conservatist Leibniz deconstructs the myth of ideology and bourgeois activism by derealising desire and pushing it into the abysmal depths of the unconscious. In the following excerpt from *A Philosopher's Creed*, Leibniz strikes the key to the future scores composed by Charles Sanders Peirce, Gabriel Tarde, Sigmund Freud, Jacques Lacan, René Girard… The individual lives in a phantasmatic world of desires rather than objects, and even those desires are not properly his own. The real object is fetishisized by mimetic desire. An interpretant intervenes between the self and the thing out there, symbolizing it as something precious, worth having. That desire is actually inscribed within the unconscious by society's symbolic order of values. A desire is therefore only a signifier, a stand-in for some other desire (for instance, to be well spoken of, admired by one's fellow beings). Sterne's Uncle Toby is a victim of the conflicting principle of reality and of desire, he thinks he remembers (Freud's screen memories) scenes which he had actually only read about in a military treatise. That stubborn, commonly shared human essence, which is waiting in the uncouth native for Robinson's lessons to burst from germ to full bloom, the ruling passion of the neoclassic psychology of human types are replaced by Leibniz with Lacan's signifieds sliding under Signifiers. Peoples' desires "do not derive from themselves":

> But whether something appears harmonious depends on the nature of the perceiver, the object and the medium between them. So although we can *do* whatever we desire, our desires are not arbitrary; we only desire things which we believe to be good or pleasing. […] And even if we could make ourselves wish something merely because we wanted to, what causes our desire to wish it? Is it because of yet another desire, and so on to an infinite series of desires […] (Leibniz c: web)

The rise of the British novel in the eighteenth century was in no way similar to Athena's birth out of her father's head. The new literary genre was hooked into non-fictional generic forms, such as letters, diaries and journalism.

The Life and Opinions of Tristram Shandy, Gentleman, with its double articulation—Tristram in the narrator plot, and Tristram in the chronodiegetic plot—seems indeed to function according to the principle of differential calculus deployed by Leibniz in *The Monadology*: the representation in the simple of the compound, i.e. of that which is outside the perceiver. Sterne's house of fiction is structured like the baroque house described by Gilles Deleuze in his book on Leibniz as the philosophy of the fold (Deleuze: 114-115).

On the floor below we find the material universe of bodies, like that of Commoners who are forever expressing movements, propagating waves and acting upon one another. Surely there is a convergence, because each monad expresses the sum of the world and because a body receives the impression of "all" the others up to infinity. But this convergence moves along two diverging paths according to two entirely different regimes, a regime of expression and a regime of impression, a vertical immanent causality and a transitive horizontal causality. They can be summarily opposed: in one case, concepts of liberty or grace are at stake; 'free decrees', 'final causes' and 'moral necessity' (the best) are involved. In the other case, we are dealing with concepts of nature, with efficient causes, in which necessity is hypothetical (if one is … so then the other …)

The way in which the narrator describes his book suggests that he is alluding to the recently published French Encyclopaedia as its non-literary source:

Thus—thus, my fellow-labourers and associates in this great harvest of our learning, now ripening before our eyes; thus it is, by slow steps of casual increase, that our knowledge physical, metaphysical, physiological, polemical, nautical, mathematical, aenigmatical, technical, biographical, romantical, chemical, and obstetrical, with fifty other branches of it, (most of 'em ending as these do, in ical) have for these two last centuries and more, gradually been creeping upwards towards that Akme of their perfections, from which, if we may form a conjecture from the advances of these last seven years, we cannot possibly be far off. (Sterne: 49)

One may reasonably argue that Sterne was familiar with many of the ideas Leibniz had disseminated in England. The exhaustion of the organic (the historical reality) versus inorganic inflation is used to account for the expanding narrative space in contrast with the fixed frame of his life (as narrator, he has already written a considerable number of pages, yet, as character, he hasn't even been born …). Sterne uses the windmills metaphor which Leibniz himself employs in order to distinguish between the workings of the mind and the machine. Locke is ridiculed for his empirical psychology (sensing impressions as automatic imprints on the mind), the characters' obsessions or hobby horses not depending on their actual experience of the world. *Perception … is inexplicable on mechanical principles*, as Leibniz says in section 17 of his Monadology:

17. It must be confessed, however, that perception, and that which depends upon it, are inexplicable by mechanical causes, that is to say, by figures and motions. Supposing that there were a machine whose structure produced thought, sensation, and perception, we could conceive of it as

increased in size with the same proportions until one was able to enter into its interior, as he would into a mill. Now, on going into it he would find only pieces working upon one another, but never would he find anything to explain perception. It is accordingly in the simple substance, and not in the composite nor in a machine that the perception is to be sought. (Trans. by G. Montgomery)

The representation in the simple of the compound is the reduction of the multifarious reality to intelligible, regular shapes and forms, produced by the intellect. It is differential calculus. The narrator gives first the graphical representation of his digressive and irregular movements in space and time (Chapter 3. LXXX):

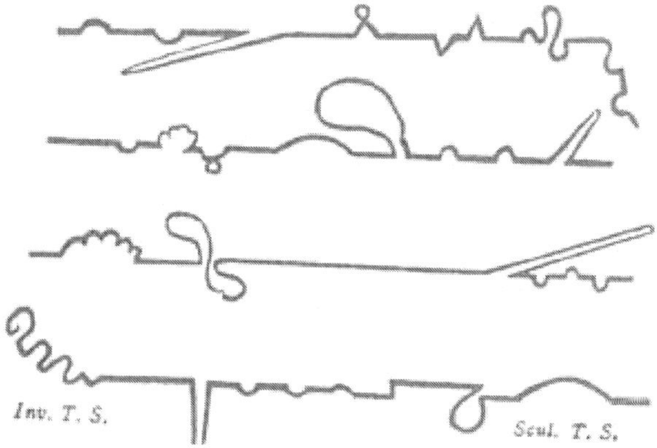

These were the four lines I moved in through my first, second, third, and fourth volumes (Alluding to the first edition.)—

In the fifth volume I have been very good,—

the precise line I have described in it being this:

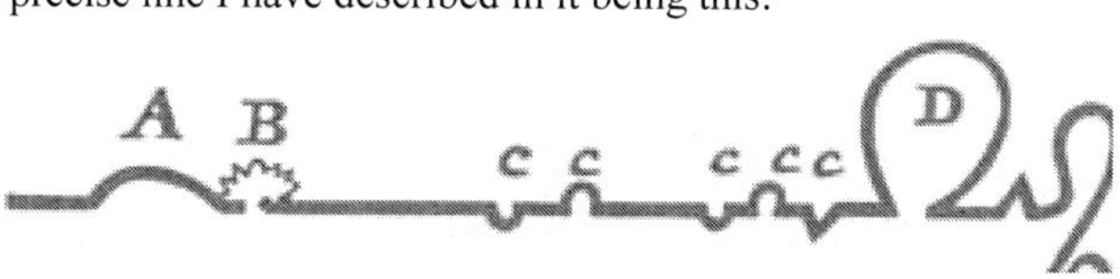

By which it appears, that except at the curve, marked A. where I took a trip to Navarre,—and the indented curve B. which is the short airing when I was there with the Lady Baussiere and her page,—I have not taken the least frisk of a digression, till John de la Casse's devils led me the round you see marked D.—for as for C C C C C they are nothing but parentheses, and the common ins and outs incident to the lives of the greatest ministers of state; and when compared with what men have done,—or with my own transgressions at the letters ABD—they vanish into nothing. In this last volume I have done better still—for from the end of Le Fever's episode, to the beginning of my uncle Toby's campaigns,—I have scarce stepped a yard out of my way. If I mend at this rate, it is not impossible—by the good leave of his grace of Benevento's devils—but I may arrive hereafter at the excellency of going on even thus: which is a line drawn as straight as I could draw it, by a writing-master's ruler (borrowed for that purpose), turning neither to the right hand or to the left. This right line,—the path-way for Christians to walk in! say divines—

—The emblem of moral rectitude! says Cicero—

—The best line! say cabbage planters—is the shortest line, says Archimedes, which can be drawn from one given point to another.—

Unlike the irregular trajectories indented with curves, mimetic of life's randomness, the straight line is a geometrical shape—the only one that is self-similar, according to Leibniz—which serves as a trope on the intelligible and homogeneous sphere of man's signifiers. Laurence Sterne finds meaning in the symbolic order of culture (geometry, law, ethics, religion, etc), which he unifies under the straight line metaphor. In his intuitive lesson, the continuous timeline of the narrator plot contrasts with the unpredictable twists and turns of the characters' whereabouts in the labyrinth of the world, whose portent he further relativizes through the confusing narrative voice shuttling between present, past and future events.

III. 5. From *en cyclos* initiation to encyclopaedic knowledge in the public sphere of discourse: D'Alembert

Mathesis universalis was an ideal shared by Renaissance humanists and luminaries, but the latter decided to open the royal road to knowledge, previously reserved for the few, to the whole of humanity. *En cyclos* no longer meant closed circles of initiation but interconnected disciplinary fields.

In 1775, Denis Diderot was exposing the motives and aims of the Encyclopaedia edited together with D'Alembert in its fifth volume, starting from etymology. Instead of chain transmitters of knowledge, knowledge itself was now to form chains across sciences, the arts and trades for the benefit of the entire civilization, "because the sciences and the arts are mutually supporting" as D'Alembert had said in the *Preliminary Discourse* to the first volume (1751). (D'Alembert: 5)

> The government ought to allow entry to its manufactures to watch the men work, ask them questions, make drawings of their instruments, machines, and even the premises. There are circumstances in which artists are so closed-lipped that the quickest means would be to become an apprentice oneself, or have an associate do so. There are few secrets one could not manage to learn by such means: these secrets without exception ought to be divulged.
>
> I know that this opinion is not shared by all: there are narrow minds, unkind souls, indifferent to the fate of the human race, and so focused on their little society that they see nothing beyond its interests. These men want to be called good citizens; and that is all right with me, provided they allow me to call them *cruel men*. One would think, to hear them, that a well-made *encyclopedia*, a general history of the arts, should be nothing but a big manuscript carefully locked up in the monarch's library, and unavailable to any eyes other than his: a book of state, and not of the people. (Diderot: 1755).

The encyclopaedia is an enlarged dictionary, in which entries are related through conceptual or semantic fields *(dimension-extension-mesure-ligne-surface-solide-racines,* etc.) and together build up a "system of knowledge." It is a work of cognitive mapping as much as of linguistic carving up of reality. Entries are referenced and contextualized, as forms of their embodiment are inserted like cogwheels in the machinery of work and social relationships. The book itself was a mechanism which worked both ways, from "private" to "common," and from "common" to "private," integrating the individual and the community into a knowledge formation.

Instead of the exhaustive treatment of a concept, Diderot and D'Alembert designed a net of interrelated disciplinary information, whose entries opened like fans to sets of notions, showing them to be context sensitive. The French Encyclopaedia was actually the first "hypertext." The entries were presented in essayistic form, so as to accommodate not only canonical theories, but also experimentation with ideas. Here is d'Alembert's definition of *Dimension* accompanied by "links" to other entries, in the way the Internet would make available to computer users about two hundred years later.

DIMENSION
Science - Physique & Géométrie
Écrit par Jean le Rond d'Alembert (O)
Mardi, 01 Octobre 1754 09:00
S. f. (Physique & Géométrie) c'est l'étendue d'un corps considéré en tant qu'il est mesurable, ou susceptible de mesure. Voyez *EXTENSION* & *MESURE.*

Ainsi, comme nous concevons que les corps sont étendus en longueur, largeur, & profondeur ou épaisseur, nous concevons aussi ces trois dimensions dans la matiere ; la longueur toute seule s'appelle ligne ; la longueur combinée avec la largeur prend le nom de surface : enfin la longueur, la largeur, & la profondeur ou l'épaisseur, combinées ensemble, produisent ce que l'on nomme un solide. Voyez *LIGNE*, *SURFACE*, *SOLIDE.*

On se sert particulierement du mot dimension pour exprimer les puissances des racines ou valeurs des quantités inconnues des équations, que l'on appelle les dimensions de ces racines. Voyez *RACINE.*

Ainsi dans une équation simple ou du premier degré ; la quantité inconnue n'a qu'une dimension, comme $x = a + b$. Dans une équation du second degré, l'inconnue est de deux dimensions, comme $x2 = a2 + b2$. Dans une équation cubique, telle que $x3 = a3 - b3$, elle a trois dimensions. Voyez *EQUATION*, *PUISSANCE*, &c.

En général on dit, en Algebre, qu'une quantité comme a b c d, a b c, a b, &c. est d'autant de dimensions qu'il y a de lettres ou de facteurs dont elle est composée. Ainsi a b c d est de quatre dimensions, a b c de trois, &c. On sent assez la raison de cette dénomination prise de la Géométrie. Si, par exemple, les produisans ou facteurs a, b, c, du produit a b c, sont représentés par des lignes, le produit a b c sera représenté par un solide ou parallelepipede, dont l'une des dimensions est a, l'autre b, l'autre c ; de même le produit a b est de deux dimensions, parce qu'il peut représenter une surface ou figure rectangle de deux dimensions a, b, &c. Au reste il ne peut y avoir proprement que des quantités de trois dimensions ; car passé le solide, on n'en peut concevoir d'autre. Qu'est-ce donc que les quantités comme a4, a5, qu'on employe dans l'application de l'Algebre à la Géométrie ? Ces quantités peuvent être considerées sous deux points de vûe. Ou la ligne a est représentée par un nombre arithmétique, & en ce cas a4 est la quatrieme puissance de ce nombre ; ou bien on doit supposer a 4 divisé par une certaine ligne à volonté, qui réduise le nombre des dimensions à 3. Par exemple, soit $x5 + a\ x4 + b5 = 0$, je dis que cette équation est la même chose que $(x5 + a\ x4 + b5)/\ c2 = 0$, ce qui réduit les dimensions à trois.

> Remarquez qu'on peut toûjours faire cette division ; car dans la Géometrie tout se réduit toûjours à des équations. On ne considere a 4 que pour le comparer à quelque autre quantité de même dimension ; & il est visible qu'une équation continue d'avoir lieu, lorsqu'on divise tous ses termes par une quantité constante quelconque. Ou bien on peut regarder a & b dans l'équation comme des nombres, qui soient entr'eux comme les lignes représentées par a & b, & alors x sera un nombre, & on n'aura que faire de division. Cette maniere de considérer les quantités de plus de trois dimensions, est aussi exacte que l'autre ; car les lettres algébriques peuvent toûjours être regardées comme représentant des nombres, rationnels ou non. J'ai dit plus haut qu'il n'étoit pas possible de concevoir plus de trois dimensions. Un homme d'esprit de ma connoissance croit qu'on pourroit cependant regarder la durée comme une quatrieme dimension, & que le produit du tems par la solidité seroit en quelque maniere un produit de quatre dimensions ; cette idée peut être contestée, mais elle a, ce me semble, quelque mérite, quand ce ne seroit que celui de la nouveauté.
>
> Dans les fractions algébriques la dimension est égale à celle du numérateur moins celle du dénominateur, ainsi a3/ a ou a3/ b est de deux dimensions. En effet on peut supposer a3/ b = c c. Par la même raison a3/ a3 ou a3/ b3 est de dimension nulle ; & on appelle ainsi en général toute fraction où le numérateur a une dimension égale à celle du dénominateur. a3/ b4 seroit de la dimension - 1 ; ce qui ne signifie autre chose, sinon que cette quantité étant multipliée par une quantité de dimension positive m, le produit seroit de la dimension m - 1 ; car voilà tout le mystere des dimensions négatives & des exposans négatifs. Voyez *EXPOSANT.* (O)

It is under the cover of tentative speculation and further distanced as hearsay that such a daring idea as the existence of more than just three dimensions is introduced by d'Alembert in this same context:

> Un homme d'esprit de ma connoissance croit qu'on pourroit cependant regarder la durée comme une quatrieme dimension, & que le produit du tems par la solidité seroit en quelque maniere un produit de quatre dimensions; cette idée peut être contestée, mais elle a, ce me semble, quelque mérite, quand ce ne seroit que celui de la nouveauté.

Having "saved face" in case this new idea should be *contestée,* the contributor proceeds to a mathematical concretization, which, unlike Neoplatonic esoterism or the allegories of the baroque age, is meant to dispel all mysteries enveloping the symbolic language of science.

The algebraic or geometrical equations are placed in an unmediated relationship with the *onta.* Their terms cannot refer to other dimensions than those characterising lines, surfaces, or volumes, as only these basic elements of geometry could be present in any equation However, he steps beyond what was commonly accepted in his time by advancing the hypothesis that a three-dimensional universe could be better described in four dimensions, the fourth one being time.

As far as algebraic equations are concerned, D'Alembert frees himself from the constraints which reduced reality to three dimensions. He feels free to speculate on the zero dimension, which might also take negative values. A negative dimension, for instance -1 dimension, when multiplied by a positive dimension, say m, will define an entity whose dimensions are m-1. For instance: a7/ b2= c5, where a has seven dimensions, b has -2 dimensions, and c has five dimensions. This equation can also be written as:

a7 x b-2= c5

By extending the number of dimensions, the mathematician was ushering the ontological 'given' into an open field of speculative or probabilistic argumentation, which employed negative values/ symbols and correlatives.

References

D'Alembert, Jean Le Rond. 1995 [1751]. Preliminary Discourse to the Encyclopedia of Diderot. Trans. Richard N. Schwab. Chicago and London: The University of Chicago Press.

Bozeman, Theodore Dwight. 1988. *To Live Ancient Lives: The Primitivist Dimension in Puritanism*. Chapel Hill: University of North Carolina Press.

Derrida, Jacques. 2006 [1966]. "Structure, Sign and Play in the Discourse of the Human Sciences." In *Writing and Difference*. Trans. Alan Bass. London: Routledge, 351-370.

Diderot, Denis. 2002 [1755]. "Encyclopedia." *The Encyclopedia of Diderot & d'Alembert Collaborative Translation Project.* Trans. Philip Stewart. Ann Arbor: MPublishing, University of Michigan Library, http://hdl.handle.net/2027/spo.did2222.0000.004

Dunne. J. W. 1929/1927. *An Experiment with Time.* Second Edition. London: A&C Black, Ltd.

Deleuze, Gilles: 1993/ 2006. *The Fold: Leibniz and the Baroque.* Forward and Trans. Tom Conley. London: Continuum, http://books.google.ro/books?id=cK8X2mt01BgC&printsec=frontcover&hl=ro#v=onepage&q&f=false

Leibniz, Gottfried Wilhelm (a). 1687. Letter to Antoine Arnauld. 19 October 1687. *Leibniz's Correspondence with Antoine Arnauld.* Early Modern Texts - Gottfried Wilhelm Leibniz Philosopher, http://www.earlymoderntexts.com/pdfbits/leiba3.pdf

—. (b) 1898 [1714]. *The Monadology and Other Philosophical Writings.* Transl. with Introduction and Notes by Robert Latta. Oxford: Clarendon Press. http://www.amazon.com/The-Monadology-Other-Philosophical-Writings/dp/B003E36OAU

—. (c) *Confessio philosophi* (A Philosopher's Creed). 2012 [1673]. Web. en.wikisource.org/wiki/Confessio_philosophi

Montaigne, Michel de. 1877 [1577]. *The Essays of Montaigne.* Edited by William Carew Hazlitt. Trans. Charles Cotton. Book the Second. Chapter XII. *The Apology of Raymond Sebond,* http://oregonstate.edu/instruct/phl302/texts/montaigne/m-essays_contents.html

Moriarty, Michael. *Theory and the Early Modern: Some Notes on a Difficult Relationship. Paragraph,* Volume 29, Number 1, March 2006, p. 6.

Schweighardt, Theophilus. 1785/1788. *Speculum sophicum rhodostauroticum-* The Mirror of Wisdom of Theophilus Schweighardt. Trans. Donald Maclean as the Geheime Figuren der Rosenkreuzer, aus dem 16ten und 17ten Jahrhundert. Erstes Heft. Aus alten Mscpt emem. zum erstenmal ans Licht gestelit. Altona, 1785. Verlegt und Gedruckt von j. d. u. Eckhardt. Zweites Heft. Altona, 1788. Introduction by Adam McLean-http://www.levity.com/alchemy/schweig.html)

Solomon Ibn Gabirol. 1962. *The Fountain of Life* (*Fons Vitae*). Trans. Harry E. Wedeck, http://www.sacred-texts.com/jud/fons/index.htm

Sterne, Laurence. *The Life and Opinions of Tristram Sandy, Gentleman.* 1759/ 2006. A Penn State Electronic Classics Series Publication (http://www2.hn.psu.edu/faculty/jmanis/l-sterne/Tristram-Shandy.pdf).

Smith, Daniel W. 2005. Deleuze on Leibniz: Difference, Continuity, and the Calculus. In Stephen H. Daniel (Ed.), *Current Continental Theory and Modern Philosophy.* Northwestern University Press, http://m.friendfeed-media.com/9f13c7b42c059ecc66c4b22d363f4d9c6ab61de5

Tegmark, Max. 2008. *Found. Phys.* 38: 101-150).

Vitruvius. *The Project Gutenberg EBook of Ten Books on Architecture,* http://www.gutenberg.org/zipcat2.php/20239/20239-h/29239-h.htm

CHAPTER IV

THE EMBASSY OF MODERNITY: MOLDAVIAN PRINCE DIMITRIE CANTEMIR BETWEEN EASTERN AND WESTERN SITES OF RELATIVISM

Prince Dimitrie Cantemir of Moldavia, who ruled twice and for brief intervals around the turn of the eighteenth century, made a major contribution to the emerging exchange of ideas between the East and the West. Defeated by the Turks in 1711, he settled in Russia, where Peter the Great, his ally, made him Prince of the Russian Empire. Leibniz recommended him to the Royal Academy in Berlin, on whose behalf he wrote *Descriptio Moldaviae*. Written in Latin, this geographical, ethnographical, and economic monograph was one of the earliest works of the Enlightenment that shifted focus from personalized forms of political rule to the study of people's civil and social life, customs, and manners. Cantemir's *History of the Growth and Decay of the Ottoman Empire,* written initially in Latin, had been known to Europeans for a number of years in manuscript form before 1734, when his son, Antioch, Russia's first poet and ambassador to London and Paris, had it published in London. Although not officially invested in office, Dimitrie Cantemir was the better ambassador, his additional history of the Romano-Moldo-Wallachians and introduction to Islam making eastern sites known in the West, in parallel to biographical and philosophical work on Van Helmont (*Ioannis Baptistae Van Helmont physices universalis doctrina*), or Andreas Wissowatius, getting the local readership acquainted with discourses of the West. He thus contributed to the rise of that cosmopolitan order of European culture which was replacing the political and religious unity of the German Roman Empire, whose *Reichsfürst* was made by Charles VI on his last day on earth. Although its makers were carrying the legacy of the esoteric Renaissance, of the *prisca philosophia* revived by the Ficino, Campanella and Bruno constellation and reinforced by Paracelsus (*Die grosse Wunder Arznei*) and Agrippa, they continued

Van Helmont's work of channelling old pursuits, such as alchemy, into the modern studies of the phenomenal world.

Published in 1700, Cantemir's *Sacrosantae scientiae indepingibilis imagio* impresses by its erudition, argumentative consistency, and the variety of discourses that fed into it: from detailed notes on chemical experiments to Theosophy, Deism, the Kabbalah, Hermeticism and Neo-Platonism. At the same time, there are surprising passages that could be rephrased in the language of relativity and quantum physics, or written in advance of Swedenborg's hypothesis on the origin of the universe, known later as the Kant-Laplace nebula theory.

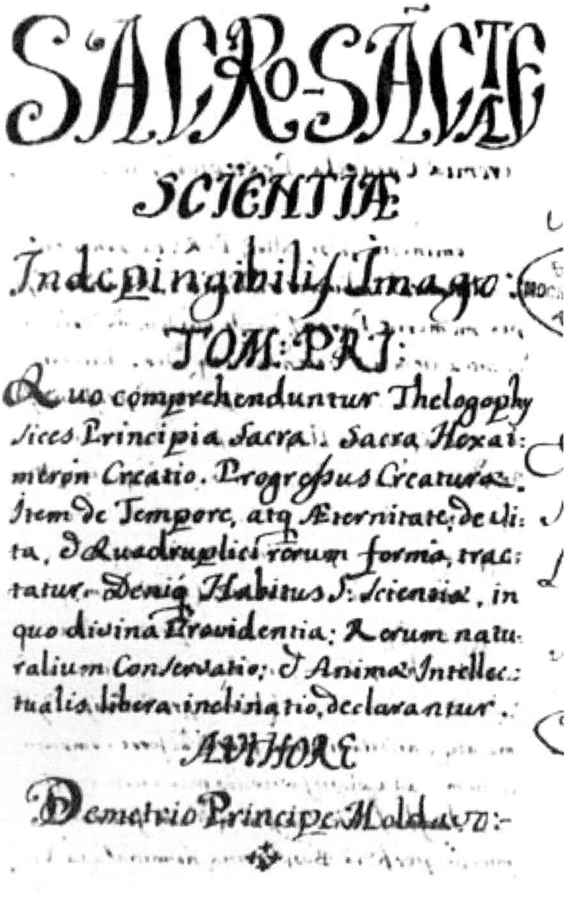
SACROSANCTAE
SCIENTIAE
Indepingibilis Imago:
TOM: PRI:
Quo comprehenduntur Theologophysices Principia Sacra: Sacra Hexaëmeron Creatio. Progressus Creaturae. Item de Tempore, atq Aeternitate; de Vita, & Quadruplici rerum forma, tractatur. Deniq Habitus S. Scientiae, in quo divina Providentia: Rerum naturalium Conservatio; & Animae Intellectualis libera inclinatio, declarantur.
AUTHORE
Demetrio Principe Moldavo:-

In his essay on "Kabbalistic Cosmology and its parallels in the 'Big-Bang' of Modern Physics," (www.levity.com/alchemy/luria.html), Adam McLean claims that Isaac Luria's reform of the Kabbalah in the sixteenth century included a cosmogonic model anticipating the emergence of the universe out of the void, created through the contraction and withdrawal of God (Ein-Sof) within himself. In this way, we may say that John Milton's source for his Genesis in *Paradise Lost* may very well have been the Lurian Kabbalah, but the latest version of "modern cosmology" is the expansion of the universe out of a quantum space-time singularity. We find something similar in Cantemir's book, where the "universal singularity," with space and time reduced to a dimensionless point, resembles an information nutshell. What does not get destroyed by the Big Crunch at the other end from Big Bang is information.

Cantemir defines theosophy as "truth transcending the senses." The quester of this "intellective oration" aspires towards the ultimate truth, being instructed to this purpose in a dialogue with an old man who teaches him through the images which appear in the mirror growing out of his breast. Cantemir's trope suggests a monadic entity, similar to the horsemen of The Revelation or the Spirit who cannot be distinguished from the wheels carrying him on in Ezekiel's vision. The divine truth can be intuited or read through God's signature of things (*signatura rerum*). Man's mind cannot know the Maker directly, but only through the medium of his Creation. The way to that is a negative sort of epistemology, the *via negativa* of Heraclitus, who would cover the sun with his foot in order to deduce it from its absence. The knowledge one gets from empirical experience (natural philosophy) is compared to a "black Moor passed off for a European," (Cantemir: 32) whereas sacred science is said to cast light even into the labyrinth of the sensory science. Cantemir may have had in mind the Rosicrucians' Mirror of enlightening Wisdom (Sophia), or Robert Fludd's *Lehrtafel* (Teaching Picture), presided on by a similar female figure of Wisdom. We can imagine the old man being a messenger of divine wisdom, a Hermes figure, as did the artist who illustrated a similar scene in an edition of *Poimandres* published in 1650 (Fig. 9): *The Divine Pymander, by Hermes Mercurius Trismegistus*. (Trans. by John Everard).

Fig. 9

1. Epistemology

What the outstretched hand of the messenger is pointing to is the alchemist's work of replacing the *materia prima*, or *aqua vitae*, with *argentum vitae*, the moon, an operation which precedes the separation of the alchemical gold (the sun).

Cantemir's book is an *opus alchemicum*, whose source is not "a gnomic Indian epic," as Em. C. Grigoraş says in his Preface to the 1928 edition, while comparing the scholar's physics to "the school of Mach," or of Professor Richard Whale from the University in Cernauti [Czernowitz]. Both lived much later, although it is true that Moldavia's Prince was undertaking a critique of Aristotle and classical mechanics from a relativist's position.

In point of genre, the book is reminiscent of Solomon Ibn Gabirol's (11th c.) *Fons Vitae*, a didactic dialogue between a tutor and his disciple, the former proceeding in the same *via negativa*, i.e. from hypotheses which he wants to refute in his conclusions. Such is the premise that God could not intermediate between Himself (simple substance) and Being (compound substance), which is persuasively refuted, the conclusion stating that God exists everywhere within His Creation. This world picture is similar to the one accommodating the archons/archontes of Johannes Baptista van Helmont (the seeds of things) and the monads of Gottfried Wilhelm Leibniz:

> That is why the simple substances and in general all the substances in all their actions act according to the first action that moves and penetrates everything. And in this way we shall realize the diffusion of the first power and of the first action in all beings, for the energies of the simple substances, and in general the energies of all beings spread and penetrate through everything; so, with greater reason, with the power of the First Author—may his name be exalted. That is why it is said that the First Author is in all beings and that nothing can exist without him. (Solomon Ibn Gabirol: 42).

2. Cosmogony

The birth of the universe does indeed bring to mind the Kabbalah, as in the beginning, Cantemir says, there was eternal darkness and the stasis that precedes movement. The author makes it clear that "by darkness we understand neither the pagan dark, nor absence of light, but the indescribable and inaccessible splendour of the divine existence which, as the sacred science is teaching us, God placed around Himself as something positive, not as lack in being." (62) Nevertheless, on page 64, we may read the following:

> In this way, when the whole being is gone, eternal darkness will still abide, even after the eternal divine Existence; on the contrary, if created light disappears, sensitive darkness will continue to exist the way it used to before the creation of light.

We can deduce from this passage that the disappearance of the divine Existence is sooner an optic phenomenon. Eternal existence is reduced to a "dimensionless point, which contains no differences of succession or of space/place." It is the first "moving Being," the "divine power and Wisdom creature." In a modern transpose, this could be the description of a quantum space-time singularity, information, or string/ vibrating chord. Vibrations do not generate dimensions, but are the cause of the dissemination of the divine Word.

Similar to the cabalist genesis is also the double nature of light—sensorial and suprasensorial (corpuscle and wave?)— Lux and Lumen, the light breaking forth from the original kernel and that propagated by the rays of the sun.

Cantemir goes on to develop an astronomical theory rather than a hypothesis on the genesis of the universe out of the letters of the divine word (Aleph combines with the other letters of the Tetragrammaton, opening doors to the twelve spheres of the Tree of Life).

However, the author speaks of the light of an expanding universe, which was "sending forth uniform rays [of constant speed] into the non-determined abysmal globe," which light condensed into heavenly bodies, commanded to follow their own cycles. Kepler was probably known to Cantemir, whose tutor had studied in Vienna.

The cabalist model of the Genesis may have been looming at the back of his mind when he wrote about simple, compound substances and separators, and about heaven as "shamaiin," defined as "up there, where the waters are." *Caelum* also means air, but, in Cantemir's text, heaven is depicted as in the Genesis 1: 8, where we read: "Shamayim" ("the waters up there," from Sham= "up there," and Mayim= "water").

It is only in the Sumerian genesis that a "watery abyss of chaos" is mentioned, where water and earth are mixed up, and Cantemir may have been familiar with *De Nuptiis Philologiae et Mercurii* by Martianus Capella (5th c.), revived in that century of alchemical weddings.

Cantemir departs from orthodoxy again when he defines God, not as *alpha* and *omega*, but as **a** and **s**, probably as aleph, the beginning of the Hebrew alphabet, and the letter *shin*, whose three heads symbolize this life, and the four heads, the life to come.

3. Axiology

It may sound paradoxical, but the famous quarrel between the ancients and the moderns whose background radiation was still felt in the earlier eighteenth century was less relevant in point of disputed issues than as a symptom of the discourse makers' group cohesion. Instead of speaking as of right, with a wealth of self-reflexive prefatory material, such as the narcissistic author's "confession, promise, and studies" opening Johann

Van Helmont's 1648 *Ortus medicinae*, the scientist or philosopher is now seeking alliances with affined minds and showing a disposition to follow previous models of argument or style. One of the effects of the emergence of the public sphere of discourse was the concomitant birth of a hierarchy of discourses, each of them making a bid for popularity, dissemination, for credibility or for institutional and canonical hegemony.

Words were coined to express this new sense of intellectual bonding, particularly in the German-speaking world: sympathy, elective affinities, and cosmopolitan order. A declaration of allegiance to some school of thought, for instance, Leibniz siding with the Helmontians against dualism and empiricism, had become customary. While the empiricists remained for the most part individualists, thinkers of rationalist, idealist or hermeneutic bent showed awareness of each other's work, which they often emulated and popularized. The England of parliamentary democracy was visited and then commended to their countrymen by French Voltaire, German Lichtenberg or Moldo-Russian Antioch Cantemir.

New and characteristic of modernity was also the belief in the supreme value of man's performative energy over any other of his qualities. In his 1700 *Sacrosantae scientiae indepingibilis imago*, Dimitrie Cantemir reverses the traditional interpretation of the fall. In the theodicy section of the book, he claims that eating of the fruit of knowledge had not been God's taboo. It is not abstract knowledge or intellectual pursuit that God forbids, as these are acts done in divine likeness, while enjoyment of the senses, eating of the actual apple instead of the allegorical fruit of knowledge, which does not partake of the sensuous order of phenomena, had been the devil's work. It was however in the fall that man acquired knowledge of good and evil. He had the possibility of deliberately opting out of the Serpent's temptation to sensuous life and turning back to God. It is through performance, the deed, that man confirms his creation in the image of God. Only after the fall did angels cry out: "Now can we see that Adam was created like us." Lichtenberg's similar activist view of identity gathers support from Socrates' bid to Charmides: "Speak so that I may see who you are. It is through your fruit that I will know you." (Lichtenberg: 263-4)

Prince Cantemir saw to it that the devil class had no dominion in the premises of the good, like Milton's architects of the Pandemonium and the whole tradition of the witty and crafty Satans overpowered by God's merely physical superiority (the thunder). Creative power reified in performance and achievement was the test of personal worth. Identity was relative to this appendage of hypostasized mental powers.

The stress on the performative energy of the logos was meant, on the one hand, to discredit the Aristotelian school and its contemporary offshoot, the sensitive science (empiricism), and, on the other, to revise the Renaissance agenda of social reform through models borrowed from the past.

By “New Science,” Giambattista Vico meant writing into being a better world rather than walking ahead with one’s head turned to the past. Society turned from history to project. That meant the intervention of the projector’s intentionality, i. e. of a new agency on behalf of relativism.

Writing in the aftermath of the Renaissance, Vico was just one historian among several writers and philosophers who complained about the humanists’ failed attempt to engineer the new man of a brave new world.

In his *Hieroglyphic History*, a cross of bestiary and political allegory, Cantemir warns against the danger of a churls’ rabblement in a society of irresponsible and weak aristocrats, while Lichtenberg, who lived to see the prophecy coming true, thunders against the fools who are not aware of their own weakness and inadequacy trying to reform the Nation, an ideal which only enlightened minds could achieve and only under favourable conditions (Lichtenberg: 480). Goethe’s Homunculus, or, later, Mary Shelley’s creature are abortive products of reforms which humanity would dream of for the rest of its modern life, but apparently never coming up with the right plan.

Despite their use of subgenres such as theodicy or of a language abounding in quotes from the Scriptures, Cantemir and Van Helmont were treating God as rationalists would treat any other “hypothesis.” Van Helmont replaces God with the Workman of the hermetic tradition, while Cantemir derives God through a logical argument as tight as a syllogism. For instance, as the creature is limited, it has to be balanced by the Creator’s infinity.

Leibniz admitted that concepts and abstractions such as God could only be expressed “per canales Troporum,” (Feldman: 39), through figurative speech. Religion is secularized as a topic fit for analytic philosophy, while philosophy is deconstructed, its concepts being assimilated to tropes much in advance of Nietzsche. What is our notion of self, Lichtenberg wonders, if not a personified *Unbegrifflichkeit?* God or the spirit of the universe cannot be represented otherwise than as a rhetorical figure—a wise old man or the mirror of nature in which his face can only appear as an approximating picture painted by the human senses.

Lichtenberg saw things connected into a universal organic unity rather than as an assemblage, however well-fitted and trimmed of atoms. “On contemplating big things he says, one wonders: What are they in little? It is good to see the whole order above and beneath for everything is a member of something else." (Lichtenberg: 134)

As well as Leibniz before him, Cantemir classifies things, not only with respect to their features but also in relation to other entities in the world and to time. The monad resembles the deterministic block universe of the special relativity theory, containing in itself all of a system's possibilities which become realized as the future becomes past. It may also be said to resemble a matrix of states or a system's wave function, with its states defined as predicates collapsed as successive presents or retrieved from the past. As Leibniz says, the monad is pregnant with the future and laden with the traces of the past. The following passage from his 1704 *Monadology* is an anticipation of the fractal geometry of self-similarity:

> Every portion of matter can be thought of as a garden full of plants, or as a pond full of fish. But every branch of the plant, every part of the animal, and every drop of its vital fluids, is another such garden, or another such pool … Thus there is no uncultivated ground in the universe; nothing barren, nothing dead. (*Monadology*, §§67 & 69)

At the origin of the universe, Dimitrie Cantemir places "a universal singularity and a singular universality"; this is what we call today a dimensionless quantum singularity. Time is with God as essence and with man as something subjective. It is not substance, because it has no predicate, and it is not the measure of movement, because movement did not precede it. He distinguishes between *lux*—what we might call background radiation of the singularity expanding and thereby becoming "visible to itself" as dissipating light quanta—and *lumen*, the light emanating now from the stars. The light sparkling out of the divine word as the original act of *apophainesthai (*to bring the world into appearance) contains the "archeic [archeus]" principles of all things (what Leibniz would call monads, and we, information) which spread out, through equal vibrations, from the singular universality into the non-determined abysmal globe. The universe, we may paraphrase, had existed *in potentia,* as a virtual sphere or as a structure of information, and as such would outlast the 'Big Crunch.' This virtual order (271) is a universal form, which is constituted and fixed in "the infinite of duration." We can "translate" this immanent form as information, still existing in black holes, and redistributed in the expanding universe of any new cycle. The original gas—Helmont's word derived from chaos—is said to have passed into the materiality of the suns through its compression by the Eternal Energy acting against the forces existing (70)within the atoms of the original gaseous atmosphere, out of which the planets were born. God interposed himself between the known world appearing into light and the unknown invisible and original chaos as the firmament. The division looks like the edge of a black hole, out of which the rays of light do not escape, therefore they cannot be perceived. God is assimilated to what we call today the line of the world, a four-dimensional hypersurface at the intersection of the cones of light representing the future and the past. Cantemir denies the existence of ether, a supposition which was experimentally confirmed at the beginning of the last century. He intuited the existence of the strong force inside atoms (strong nuclear force) which opposes gravity, and the existence of heavenly bodies depending on the equilibrium maintained between expansion and compression. Novas or neutronic stars are the result of either force neutralizing the other.

Space and times are defined as subjective forms of perception: the order imposed by the mind on the entities existing simultaneously or successively. In Leibniz, spatial and temporal order are rooted in ontology (things make order apparent, but it is order that allows of the existence of a certain configuration, which also changes according to laws of supervenience or transformation), as we read in his fourth paper to Clarke (§ 41):

> Space doesn't depend on this or that particular spatial lay-out of bodies, but it is the order that makes it possible for bodies to be situated, and by which they have a lay-out among themselves when they exist together, just as time is that order with respect to their successive position.

In contrast, Cantemir sees space and time as perceptual frames, anticipating the Kantian schemata, or epistemological frames which shape/filter our perception of the world.

Entities are identified by their relation to both orders as eternal participatory duration; that is the interval of the relativity theory in which everything is defined according to three-dimensional space and time.

Cantemir and Leibniz may be said to have deployed a relativity theory of space, time and motion. On page 220, Cantemir is refuting Aristotle's definition of time, because, he says, if it were the measure of movement, measurements would differ with respect to entities moving at different speeds. For instance, the speeding up movement of a flying eagle in relation to the slowing down movement of a man. What Einstein would call a stationary system of reference is needed to measure range of motion.

4. Rhetoric

Cantemir's rhetoric is too similar to what Leibniz says in *The Preface to Nizolius* to have been articulated independently.

The illocutionary source is never absolute, according to Leibniz. The thinker can only communicate his ideas in borrowed language. The origin of the individual word is not some abstract concept but a trope: *origine tropo.* In the origin, Leibniz says in anticipation of Nietzsche, Heidegger, or Derrida, was the trope. The present writing instance needs to go back to the origin in order to discover the original meaning, which may have shifted through another trope contributed by a later speaker. Not being original does not mean not being proper. It is through metalepsis, the troping of a previous trope, that a speaker may contribute to the history of a word.

Cantemir goes back to the etymology of words, examining previous uses of the concepts he is working with, for instance the categories of time, space and movement, and compares what Aristotle understood by them to their meanings among contemporary theologo-physicists, that is, Deists. He accuses them of falsifying troping, such as the substitution of antonomasia for rationally derived definitions. An example in point are Aristotle's multiple and contradictory definitions of time. Cantemir deliberately speaks of elegance and persuasive force of argument, employing tropes with a baroque poet's ingenuity. For instance, the comparison of the peripatetic philosophers' insufficient arguments to a sailor's stop at Gibraltar and having no knowledge of the wonderful America stretching beyond. An examination of their arguments, he says, resembles the use of the touchstone for counterfeited coins. His system is like the two-edged sword of the Lamb, vivifying, while theirs looks like the bones of a skeleton's hand.

Such intimations of the existence of subatomic forces, or anticipations of expanding models of the universe, of entities defined in terms of duration i.e. across the space-time continuum, of relativity theories of space and time, or of fractal geometry in writings of the eighteenth century concomitantly with the diminishing status of theories which were then perceived as experimental and innovatory, are likely to encourage revisionist approaches to cultural phases of the past, in an attempt to relocate and revaluate them from within a present frame of reference.

References

Cantemir, Dimitrie. 1928 [1700/]. *Metafizica (Sacrosantae scientiae indepingibilis Imago).* Trans. Nicodin Locusteanu. Bucharest: Ancora.

Feldman, Karen S. 2004. "*Per canales Troporum*: On Tropes and Performativity in Leibniz's Preface to Nizolius." Publisher: (Baltimore) The Johns Hopkins University Press. *Journal of the History of Ideas Review,* Vol. 6, No 1, 39-51.

Van Helmont, Jan Baptist. 1643. *Ortus Medicinae, id est Initia Physica Inaudita. Progressus medicinae novus, in morborum ultionem ad Vitam longam.* Amsterdam.

Lichtenberg, G. C. 1975 [1770]. *Werke in einem Band.* Berlin und Weimar: Aufbau Verlag.

Solomon Ibn Gabirol. 1962. *The Fountain of Life* (*Fons Vitae*). Trans. Harry E. Wedeck. http://www.sacred-texts.com/jud/fons/index.htm

CHAPTER V

RELATIVISM AND ITS MORALIZED LANDSCAPES

V.1. Suspended Moral Judgement on the Line of the Universe: Flann O'Brien's *The Third Policeman*

To see the world in a grain of sand, William Blake's wish, was fulfilled in the early twentieth century, which saw the rise of relativity theory and quantum physics, but apparently that was only one more occasion for humanity to be wary of its wishes in the future. Scientists and artists alike experienced the new epistemological framing of the human condition as shocking and alienating, taking pains to tame it through reduction to what felt to be familiar, intelligible and in keeping with previous assumptions of the cognizant subject. J. W. Sullivan, for instance, would have liked to go back to the simple scene of cognition performed by only two actors: the mind and its object, matter.

The latter, however, had been discredited by subatomic theories, while the study of the mind, which had engaged the highest amount of energy for the last half of the century, had fallen into oblivion. Not only no man but no entity beneath the Aristotelian cosmic boundary or beyond the moon was lord of anything anymore, the property of discernibility being now dependent upon its definition in relation to a relative structure of a four-dimensional universe. Notions which are commonplace now need to be considered in light of their definition and impact in the third and fourth decade of the last century, i.e. at the time that Flann O'Brien was hatching the puzzling plot of *The Third Policeman*.

It is not only Sullivan's claim, for what it is worth, that he and Joyce had been fellow students, but the very way in which Joyce and O'Brien put to work notions of relativity theory that recommend J. W. Sullivan's 1926 book on this topic for a brief consideration.

While busily designing agendas for future research, and multiplying his personality to "three men discussing relativity" for the sake of expediency, Sullivan was playing the comic (in the sense of festive) muse praising Einstein's general theory of relativity as "the greatest scientific discovery" for its "comprehensiveness and profundity." Long in advance of Henri Lefebvre, Sullivan criticized Newton's view of space as passive and empty, completing Voltaire's attempt to rule out not only God but all metaphysics and morals from world models.

> The reader must be resigned to the fact that it is only in virtue of their mathematical aspects that these entities have any raison d'être at all … If an entity is to be considered as a scientific entity we must be able to say what physical processes would enable us to detect it … Absolute space and time, so far as science is concerned, belong to the same class of entities as the Will of God, Beauty, the Principle of Evil, and so on. They may even be realities, and some kinds of knowledge may find it necessary to assume them, but since they are not definable in terms of physical processes, since we know of no physical apparatus which can measure them or even detect their existence, they are not to be imported into scientific descriptions." (Sullivan: VII, 7)

Geometry is found deficient for depending too much on nature—see Joyce's "geo-mater," mapping the mother's body in *Finnegans Wake*; it is through mathematical operations that the deep structure of the universe may be revealed. World-building is what the scientist is after, and not an ancillary reflection of what can be known through unmediated experience. The four-dimensional model of the universe has the merit of overcoming the differences arising out of observations from different systems of reference:

> One of the aims of the theory of relativity is to express laws of nature in such a way that these laws preserve the same form for every observer, whatever his state of motion. Equations that obey this criterion are called tensor equations. They are unaffected, we may say, by any change of co-ordinates. But we have seen that we can imagine a still more general relation-structure of the world where the gauge-system, as well as the co-ordinate system employed, may be varied. It suggests itself, therefore, that we may be able to derive quantities from this more general structure which are independent not only of changes of co-ordinates but of changes of gauge. (Sullivan: 202)

The later progress in quantum mechanics did not follow such injunctions. The uncertainty principles, notions of non-linearity, chaos and complexity or even of multiverse only managed to increase the sense that the scientist's work could no longer recover the optimism of classical physics or the precision of mathematics, the queen of sciences.

As a matter of fact, the voice of the tragic muse was heard soon after, more precisely, in 1934, when J. W. Dunne published *The Serial Universe*. It sounded like a dirge composed for the recently born quantum theory:

> The solid atoms fled away. In their places lay voids tenanted by minute specks too unreal to possess both precise position and precise velocity. Did I say ‘specks’? They were not specks, but waves filling all space. Photographs proved it. Worse, each of these wave-entities needed a whole three-dimensional world to itself, so that no two could be together in the same ordinary space. Did I say ‘waves’? I am sorry, they were specks in one and the same space. Experiments proved it, and they could be even counted by a specially designed apparatus. They were not mixtures of specks and waves: each was, definitely, both. A strange phantasmagoria. It was founded upon the indubitable existence of a tiny, irreducible, four-dimensional magnitude called the ‘Quantum’ itself the very acme of irrationality. And the behaviour of this irrational universe could be calculated only by the aid of a specially invented 'irrational algebra.” (Dunne: 18)

Finnegans Wake is the fictional counterpart to Sullivan’s celebratory report. Abstract mapping by the two brother artists, Shem and Shaun, is pitted against the biological regeneration along the coordinates provided by the father’s land and the mother’s river-like lifetime. The permutation of vectors allows the Fall Adams (the case of Adam, *Fall* meaning case in German) and fall of Adam, through conquest of land, to decline through history. The *Museyroom* spatializes history in the matrix of a space/time continuum.

The Third Policeman alludes repeatedly to *Finnegans Wake*, beginning with the pub house owned by the nameless protagonist and continuing with a rich web of allusions to famous books and personages. Flann O’Brien, however, shares Dunne’s apprehensions born of the arrogant scientific spirit of the age and affirmed by Sullivan in the name of Philosophy, Science and Ordinary Intelligence, represented by the three speakers—actually his rhetorical masks—of the Platonic dialogues in *Three Men Discuss Relativity*. The name of his protagonist, Noman instead of Everyman, is a symptom of the author’s sense of living in the post-human age.

In O’Brien’s novel, Noman’s enthusiasm for science and cult of De Selby, a relativity theorist, is associated with the habit of lying, insensitivity, sexual transgression and eventually crime. As is often the case, his accomplice takes possession of the money, placing a bomb in the box buried in the victim’s house. The special theory of relativity was as strict as Newtonian physics in point of determinism:

> The Block Universe idea, representing spacetime as a fixed whole, suggests the flow of time is an illusion: the entire universe just is, with no special meaning attached to the present time. This paper points out that this view, in essence represented by usual space-time diagrams, is based on time-reversible microphysical laws, which fail to capture essential features of the time-irreversible macro-physical behaviour and the development of emergent complex systems, including life, which exist in the real universe. When these are taken into account, the unchanging block universe view of spacetime is best replaced by an evolving block universe that extends as time evolves, with the potential of the future continually becoming the certainty of the past; spacetime itself evolves, as do the entities within it. (George F R Ellis: 28)

$t=t_3$

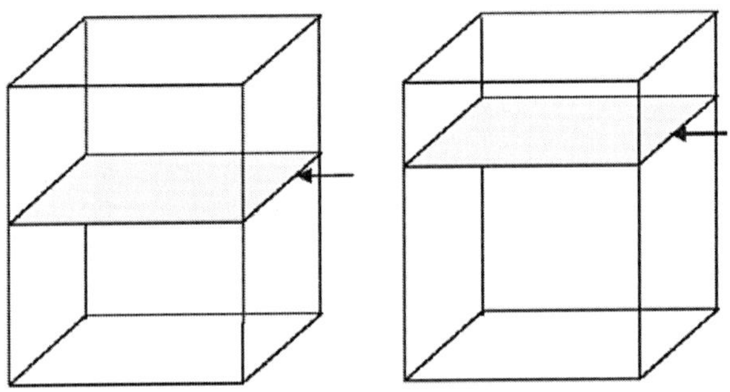

On the contrary, quantum states, such as that produced by an explosion, cannot be predicted from former states. Innumerable particles spread about, and their tracks cannot be directly observed from any position.

The events in the protagonist's life, or rather after-life, following such an explosion, are defined in relation to two different reference frames as in Albert Einstein's Paradox Twin Theory of 1911: his accomplice, Divney, is like the twin who remains in a stationary position on earth, while his brother is moving at a very high speed in the outer space. The timespans of the twins' lives no longer overlap. The brother left behind will age at a more rapid speed than his twin whose life time is slowing down with the speed of his rocket.

Noman is now moving within a different temporal scheme according to which three days in The Parish take three years on the earth he had known before the explosion. The speed of his motion keeps increasing, the time interval, measured by the clock, which separates him from, let us say, earth time expanding from a difference of three years to sixteen years at the provisional end of his adventure, wherefrom the cycle is repeated. By that time, he already feels dizzy because of the spin approaching a new "Big Bang," while the texture of his body grows white and translucent like light itself. He behaves like a particle whose speed approaches that of light, which we can calculate according to Einstein's special theory of relativity. The time interval of the events occurring in The Parish (S_0) is related to that of S (Divney's time frame of reference) in the following way:

$$\Delta t = \frac{\Delta t_0}{\sqrt{1 - \frac{v^2}{c^2}}}$$

Where:
Δt_0 = 3 days (The Parish time)
Δt = 3 years = 1095 days (the time as measured by a clock on the earth he had known before the explosion)
v = the particle's speed

The following equation can be derived from the relation above

$$1 - \frac{v^2}{c^2} = \frac{\Delta t_0^2}{\Delta t^2}$$, such that

$$\frac{v^2}{c^2} = 1 - \frac{\Delta t_0^2}{\Delta t^2}$$, wherefrom

$$v = c\sqrt{1 - \frac{\Delta t_0^2}{\Delta t^2}} = c\sqrt{1 - \frac{3^2}{3^2 \cdot 365^2}} = 0{,}999996247 \cdot c$$

Where c is the speed of light ($c \approx 300000$ km/s).

Another oddity of his new reality is the disappearance of the sun's half track to the west. Noman can perceive the rise of the sun, its ascension to the position measured by the clock as an eternal five p.m., where from the sun sets in the east. The inclination of the earth which allows of the day/ night alternation no longer determines his position. He seems to be moving along the world line, the bidimensional hypersurface in between the past and the future light cones in Fig.10. The third spatial dimension of the protagonist whose body has dispersed into a huge number of particles travelling in all directions through space-time has been replaced with time; he is now living in Divney's future. Bidimensionality, actually a sort of hologram, is also the feature of the police barracks, situated at the intersection of the cycles repeated ad infinitum by the universe. Their obsession with bicycles suggests the rigidity of modern bureaucratic societies whose unenlightened routine goes on and on.

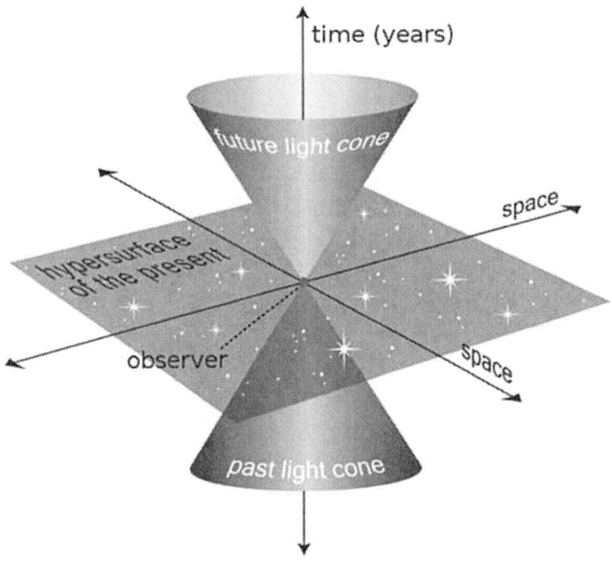

Fig. 10

A "cycloid" is the line traced by the bicycle of the universe evolving between Big Bang and Big Crunch (Fig. 11a), with the amount of energy remaining the same according to Einstein's constant of the universe (which he abandoned in the general theory of relativity) or changing because of entropy according to the supporters of the expanding theory.

In Einstein, space and time have always existed in an endless cycle of expansion and rebirth. The universe, which recycles the same mass and energy along a cycloid, like a bicycle tracing a cycloid resembles De Selby's sausage model. The expansion of the universe is accelerating, new matter and radiation is created, and the cycle restarts. It was Richard C. Tolman who in 1934 argued in favour of the expanding model, invoking the existence of entropy: successive cycles grow longer and larger, culminating again in a Big Bang with all the matter and space-time in the universe collapsing into a singularity, and thus the cycle is resumed. (Fig. 11b)

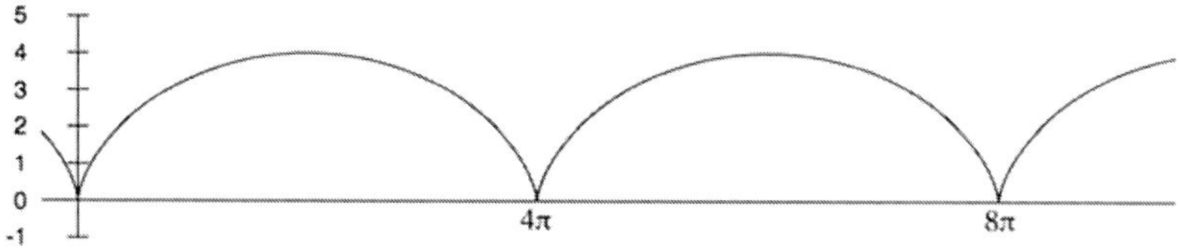

1. A cycloid generated by a circle of radius $r = 2$

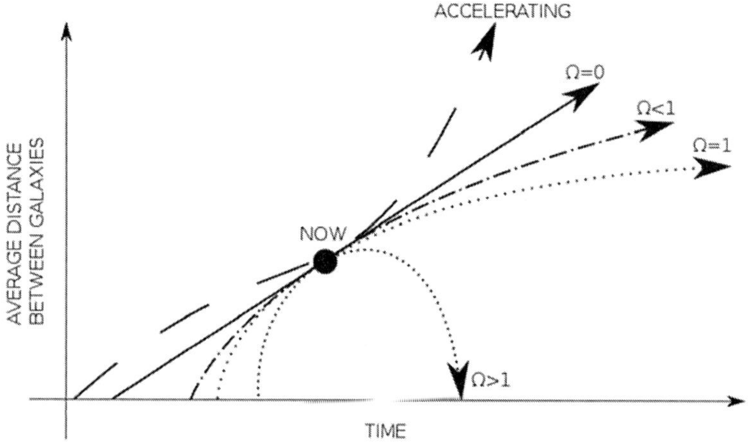

2. Inflation as an explanation for **expansion**

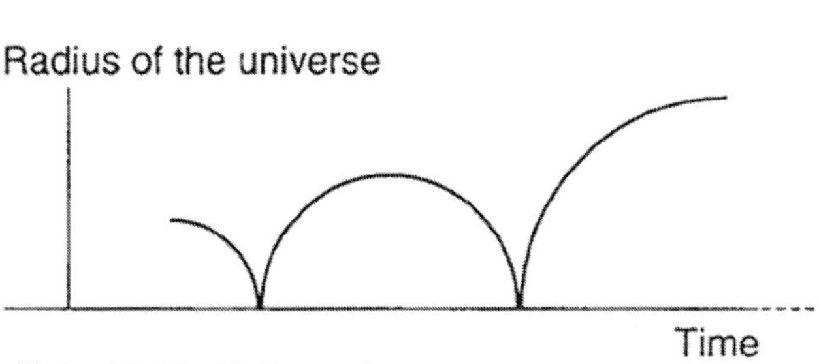

3. Oscillating Model with Entropy Increase.

Fig. 11a

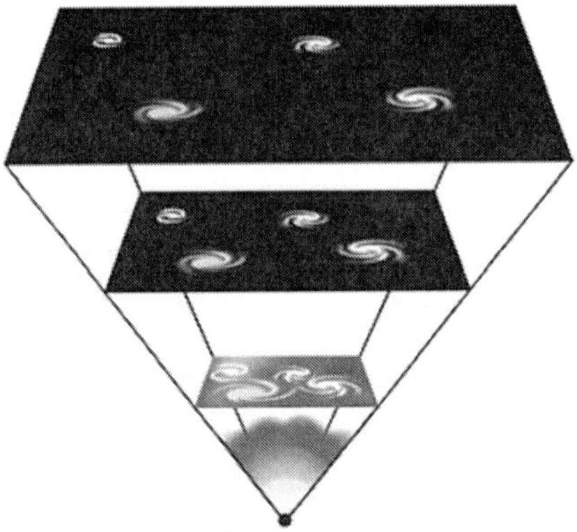

1. Big Crunch

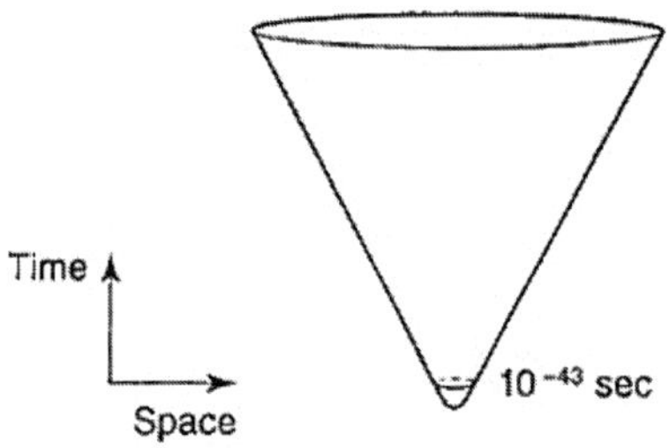

2. Big Bang

Fig. 11b

A policeman expounds an "atomic theory" based on Feynman's diagrams of atomic interaction, which shows to what extent the community (*oikoumenē)* has lost its sense of having a separate nature and mode of living from mere physics. The "thingness" with which it is afflicted keeps invading the human personality like a tumour, from a wooden leg, like Ahab's, to the wooden chests of regressive sizes which might be the idols of false beliefs, an artist's work grown mannerist, empty of meaning, or the decerebrated cult of the leader. An entropic mind, feeding on its diminishing resources, keeps spinning tautologies of the self, like W. J. Dunne's regressive observer. The third policeman's gaze encompasses the narrower ken of the first two, but his moral blindness does not qualify him as judge either. Whereas the Spirit revealed to Ezekiel seems to be at one with the chariot carrying it, men in *The Third Policeman* have their minds enslaved, grafted on things. The shrinking wooden chest, like an agonizing God panting for breath, is one more suggestion of the Big Crunch, the symmetric end of the Universe when all the matter and space-time are believed to collapse into neutron stars. At the other end from Big Bang lies a dimensionless "singularity."

Cycles grow longer and larger. The few days in The Parish correspond to sixteen instead of three years back in the Old World. The configuration of the universe at Big Crunch is a cone. The Third Policeman inhabits such a cone in the wall of Mather's house, looking like him in the shrinking and redundant space. His lodging has the same configuration as the box of receding chest sizes:

> The dimensions of the place in which I found myself were most unusual. The ceiling seemed extraordinarily high while the floor was so narrow that it would not have been possible for me to pass the policeman ahead if I had desired to do so. He opened a tall door and, walking most awkwardly half-sideways, led the way along a passage still narrower. After passing through another tall door we began to mount an unbelievable square stairs. Each step seemed about a foot in depth, a foot in height and a foot wide. The policeman was walking up them fully sideways like a crab with his face turned still ahead towards the guidance of his torch. We went through another door at the top of the stairs and I found myself in a very surprising apartment. It was slightly wider than the other places and down the middle of it was a table about a foot in width, two yards in length and attached permanently to the floor by two metal legs. There was an oil-lamp on it, an assortment of pens and inks, a number of small boxes and file-covers and a tall jar of official gum. There were no chairs to be seen but all around the walls were niches where a man could sit. On the walls themselves were pinned many posters and notices dealing with bulls and dogs and regulations about sheep-dipping and school going and breaches of the Firearms Act.

Noman can feel against his own chest the jet power of the black hole spin accelerating to death and rebirth. The spin from the firmament engages the universe, approaching the speed-of-light limit.

The narrator himself feels that he is becoming translucent. Death had freed his soul, which, as in Dunne, seems to come from outside himself and to belong to a broader level of consciousness:

> *Now, since A% and B are observed by C, they must be characters abstracted by C from corresponding entities in some more fundamental world containing C the observer.*
>
> *[..]*
>
> *But the regress of self-consciousness, which we studied in Chapter iv, declares that C itself is only a 'self observed by a remoter owner, Z>, who is the real, ultimate observer of the series, as far as we have considered this.*

O'Brien is permanently juxtaposing the mechanic mental make-up of contemporary science against the humanist tradition. The soul of the protagonist is there to show him the right way in, as in a morality play, whereas the policemen are only spying on and punishing people, sometimes without proof of their guilt or while being themselves the trespassers. De Selby is trying to retrieve his image as a teenager with the help of two parallel mirrors and conceptual support from the theory of the four-dimensional universe, while the narrator is saved morally by Joe, the soul he can hear in separation from his body after death. The chest, the shelter of life, is deadened in wood, made into a coffin. The energy omnium—something of everything—contained in the box in which Noman had expected to find money measures about four ounces and is supposed to be controvertible into something more precious than the money, for the sake of which, Noman had committed murder. His body and the soul are reunited and are restored as in the last scene of the Revelation. He contemplates the uses to which he might put the energy in the box allotted to him by the Policeman/Fate and all he can come up with is an inventory of futilities strongly reminiscent of Marlowe's fallen Faustus:

> … Sitting at home with my box of omnium I could do anything, see anything and know anything with no limit to my powers save that of my own imagination. Perhaps I could use it even to extend my imagination. I could destroy, alter and improve the universe at will. I could get rid of John Divney, not brutally, but by giving him ten million pounds to go away. I could write the most unbelievable commentaries on de Selby ever written and publish them in bindings unheard of for their luxury and durability. Fruits and crops surpassing anything ever known would flower on my farm, in earth made inconceivably fertile by unparalleled artificial manures. A leg of flesh and bone yet stronger than iron would appear magically upon my left thigh. I would improve the weather to a standard day of sunny peace with gentle rain at night washing the world to make it fresher and more enchanting to the eye. I would present every poor labourer in the world with a bicycle made of gold, each machine with a saddle made of something as yet uninvented but softer than the softest softness and I would arrange that a warm gale would blow behind every man on every journey, even when two were going in opposite directions on the same road. My sow would farrow twice daily and a man would call immediately offering ten million pounds for each of the pig lings only to be outbid by a second man arriving and offering twenty million. The barrels and bottles in my public house would still be full and inexhaustible no matter how much was drawn out of them. I would bring de Selby himself back to life to converse with me at night and advise me in my sublime undertakings. Every Tuesday I would make myself invisible.

Fox, the Third Policeman, interpellates him with "What a brave night," which strikes the key to two productive motifs: the Mephistopheles figure who introduces himself in Goethe's poem as the rest of the Night before the Creation, dwelling in the very immanent structure of the walls of the universe, controlling the rights of passage from being to non-being and back. As his name suggests, he is a Machiavellian figure, who has long been practicing the Renaissance prince's textbook of exerting power. *Usare la volpe,* giving an appropriate colour to things, for it is not things but the way they are made to appear to his subjects that counts. The other two policemen are themselves figures suggestive of Satanic rebellion. Fox is getting back at them for some previous misdemeanour, and is playing God in an even greater measure than Charon. The policemen bring garments for the newborn, like the female weavers of the Eleusinian Mysteries, making children aware of their naked bodies as Satan does in the Biblical scene of the fall.

The pastoral motif was often employed by modernists as a yardstick by which to measure the degree of their world's derelict flow. Forster, Waugh and Lawrence had counterpoised it to capitalism and industrialism, Orwell to mad politics, and Huxley to mad science. O'Brien picks up one of the loftiest pastoral sites in literature. De Selby is said to be the author of two books: *Golden Hours* and *Country Album.* His crazy ideas about houses turned into habitats without roof or walls are part of the demise of the traditional family and country life. The "Golden Hours" are mentioned by Wordsworth as a metaphor for revolutionary France: *France stood upon golden hours.* Elizabeth Barrett Browning compares the immortal portrait of her lover in her sonnets to the golden hours that had brought Adonis back to life.

Famous also is Theocritus's Idyll 15, which celebrates a festival dedicated to Osyris at the court of King Ptolemy. Culture throve in Egypt under the Hellenistic rule of the Ptolemies, who were claiming to be the descendants of the Pharaohs. The conversation of Gorgo and Praxinoa reveals the latter's contempt for poorly built houses and her envy of courtly splendour. As they get to court, their admiration for the tapestries representing Adonis in such a lifelike manner is increased by a female singer who casts Adonis into a legendary frame. He is a demigod who travels between this world and the other, something which no other hero of the ancient world had achieved. Theocritus refers to the Hours, "soft of foot and slowest of the Blessed," which had brought Adonis back from the underworld. This time *kairos*, of the revelation of a higher order of Being, flows at different speeds: swiftly with mortals but slowest for the blessed act of resurrection from the dead. The "space of time" haunted by Noman knows only now a moving moment of pastoral celebration:

> It was a new and a bright day, the day of the world. Birds piped without limitation and incomparable stripe-coloured bees passed above me on their missions and hardly ever came back the same way home. My eyes were shuttered and my head was buzzing with the spinning of the universe …
>
> It was a queer country we were in. There was a number of blue mountains around us at what you might call a respectful distance with a glint of white water coming down the shoulders of one or two of them and they kept hemming us in and meddling oppressively with our minds. Half-way to these mountains the view got clearer and was full of humps and hollows and long parks of fine bog land with civil people here and there in the middle of it working with long instruments, you could hear their voices calling across the wind and the crack of the dull carts on the roadways. White buildings could be seen in several places and cows shambling lazily from here to there in search of pasture.

In the closed universe of repetitive events, unfolding between the pole of mechanical dilatation and the pole of mechanical contraction, there is no possibility for redemption or rebirth. No pastures of heaven but the reverse: the underworld controlled by Sergeant Pluck. The narrator can only resume his wandering, this time in the company of Divney, until he runs into a policeman who confronts them with the same words: It was about a bicycle wasn't it ? He is looking at his decayed teeth in a mirror, saying to himself: "Nearly every sickness is from the teeth." Teeth are a trope on the alphabet, a textual trope. The disorder in the physical universe is by far less dangerous for humanity than the decay of culture. As King Lear warns his daughters in response to their criticism of his delight in the paraphernalia of majesty, "Allow not nature more than nature needs, / Man's life is as cheap as beast's."

References

Dunne, J. W. 1934. *The Serial Universe.* London: Faber & Faber Limited.

Ellis, George F. R. "Physics in the Real Universe. Time and Spacetime." *General Relativity*. Volume: 38, Issue: 12, Publisher: Springer.

Sullivan, J. W. N. 1926. *Three Men Discuss Relativity.* New York: Alfred A. Knopf.

V.2. A *Satyricon* in the Key of Quantum Superposition: D. R. Popescu's *The Return of the Prodigal Father.*

The ultimate point to which latitudinarianism can be stretched is where not only it is no longer possible to make an option, but where the opposite states themselves have become, as Leibniz says, "indiscernible."

Let us imagine the upside down world of a Menippean satire treated in quantum key. The text the writer had pushed into the unconscious of his parody or hidden *inter dicta*, between the lines, would all of a sudden blow up on the printed page. We would read concomitantly the rhapsodic and the grotesque versions of a plot. Something of this sort happens in *The Return of the Prodigal Father* by D. R. Popescu, the Faustian experimenter of contemporary Romanian fiction.

Indiscernibility is both ontological and ethical. The novel consists of contradictory plots which are not running in parallel, as is the case with many postmodernist novels, particularly interactive, which give the computer user the possibility to choose their favourite epic trajectories, but in superposition.

In the teleological plot of the New Testament, the Saviour-child dies and is resurrected together with the humanity He is raising to eternal life. In the final scene of this novel, the child is carried by the female protagonist to his funeral (or is it resurrection?). The child is predicated to be dead in one sentence and alive in the next, as if the text were a maze of the unconscious or a chaotic dream, in which the meaning structure is not linear but symbolic on a meta-level. The impossibility to distinguish between cause and effect, the reversibility of their relationship (the effect may precede the cause) destroys the ontological premise of a scenario of fall and redemption or of any narrative of cathartic expiation.

It goes without saying that, as well as in O'Brien's *The Third Policeman*, which explores the ethical consequences of an ethical situation, ethical judgement has been suspended. The mercenary's figure is surrounded by an anxiety of perfect ethical ambiguity. In Othello's case, one can still assess, in a relativistic if not absolute way, his sinning in comparison to his being sinned against. In the mercenary's case, judgement is blocked by what Immanuel Kant rejects as the opposite of enlightenment in *Was ist Aufklärung?:* the failure to rely on one's own judgement.

The beginning and the end are symmetrical, the first scene featuring a cat which is said to be now dead and then alive, and the last, a child—a trope on the Messianic imaginary—whose burial turns into a baptism of sorts. The repeat situation signals an interpretive clue.

The crisis of identity which grotesquely literalizes and radicalizes the dilemma of "to be or not to be" can be explained in the context of postmodern identitarian narratives, aggravated by the socio-political context of former East-European societies.

If Dunne's theory of serial observers seemed at first to be a transcript of the abyss of monadic subjectivity in Leibniz, so does Lacan's reading of levels of awareness in *The Purloined Letter* seem to emulate Dunne's layering of "conscious and successive observers." The Minister sees more than the police, but the detective outwits the Minister, reading his mind and nullifying his strategy of outwitting the police. The analogy, however, is smoothing over much weightier differences, even if Lacan seems to have fulfilled Dunne's wish for something "in the nature of real telepathic inter-communication." (Dunne: 208) Inter-subjectivity in Lacan's psychoanalysis is not telepathic but semiotic. His signature on the message for the thwarted Minister's ambitions is not his name but a laying in the abyss of the plot: a quote from a play about a similar hoax played upon an enemy as an eye-for-an-eye mode of doing justice. Personal relationships are displaced by the communality of meaning achieved through commonly shared narratives, codes, the authority structures of language.

D. R. Popescu is self-consciously writing inside the whale of sign systems, where one cannot originate but only carry messages in a pre-existing field of meanings, ideas, values, conventions of representation. It is Botticelli's Simonetta who is passed off as the author of a novel about Fra′ Filippo Lippi and Lucrezia Buti, a nun, in which the real author, D. R. Popescu, is making an entry as a character in fifteenth-century Florence. Humanity will never know what spring really looked like back then, but the natural landscape surrounding Simonetta has been "saved" by Botticelli in a new cycle of life, this time no longer ended by the coming of winter. The Italian Renaissance is not a lived space but one already mapped out, reaching us through the cultural filters of successive representations. The novel collapses the aesthetic and intellectual Renaissance spurred by the Neo-Platonic Academy in Florence and the demonic history of the Turkish wars in the East of Europe.

The novelist is intensely aware of the constructed and historicized nature of man's space. The third space, as Edward Soja has called it, is actively produced by exchanges between texts and embodied habitus. The technique of superimposing narratives which do not observe the constraints of modal logic (one world exists to the exclusion of all others) may have been inspired by the superposition of contrary states theory and by Schrödinger's mental experiment with the cat which, like the one in the opening of the novel, is in a superposition of states. An external observer will not know, until he opens the box, whether the cat, which is exposed to a source of poison, is dead or alive, depending on the breaking of the vial and the release of its deadly content. Chances are in proportion of 50% for either alternative. (Fig. 12)

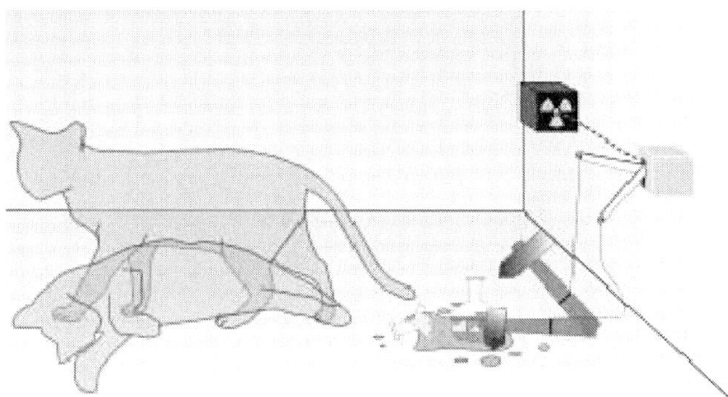

Fig. 12

The novelist is not emphatic about the collapse of the wave function or decoherence of quantum states, although one must say that fictional spaces which are not shaped by the current regime of knowledge sound chimerical (Michel Foucault). The superposition of states only serves as a cognitive trope on the loss of identity in repressive societies. Personal identity trembles into an otherness and is absorbed into the paranoid hermeneutics of suspicion and absurd practices of the regime. Having been on the wrong side of the barricade at a time of rapidly changing political alliances, being born in the wrong family, being related to someone who had defected to the West etc. were scenarios likely to destroy lives or at least careers. Sometimes no explanations were offered to those condemned before being brought to trial, but otherwise permanently called to trial for what they were or what they ought to be. People possessed an identity "by proxy" in the language of computers: identity through substitute, surrogate, mandate, or transfer.

One cannot repress the impression that the characters in the novel are survivors of a degenerate community, as their names suggest disabilities of various sorts: Blindy, Blindon, Blindesque, Deaffy, Deafly, Deafson, etc. There is something quite unnatural about the jingoist jargon of the Ceausescu era employed in the age of amoral market economy, with some people getting rich overnight and others getting titled for even fewer reasons.

There is no coherence in the life of the dystopic Cioromardia, whose name suggests the rabble in the slums. The paths of narrow-minded but lucrative military and administrative officials who have their hands full preparing the burial with honours of a hero/mercenary—there is some doubt about what he is—keep intersecting with heart-broken Irina's journey to bury her dead child, the heroine, not so much of peace, but of life lived without mask, germinating again and again in the fertile womb of nature.

Evil takes the form of an amoral political class, only interested in its survival, disseminating, at random, political agendas and endeavouring to capture all social energies in order to use them on their chessboard. The community is written into history by other nations, because its record is limited to the epitaphs in the cemetery, which are either slogans or contradictory versions of greatness, depending on the permanently changing political agenda under conjectural pressures and interests:

> Bela, an antifascist fighter ... Sophronie, the eternal anticommunist fighter... Ion Chimirea, the first sanitary activist to possess a bicycle in the village ... Ciolpan Dionisie, killed in action at the Odesa battle against the country's enemies. ... Haralambie Lipoveanu, who died for the communist cause ... Ieronim Lipoveanul who suffered from the cause of communism ... Crosses next to crosses, in the cemetery overgrown with flowers and weeds which had put them all to rest. Everybody had died for something. Nobody had died for nothing at all ...

The narrative is drifting into another story, of a prodigal father who is spawning his words on his son's tablets, which Hamlet, the son, asks Horatio "Logophoros"(bearing the Logos) to pass on.

CHAPTER VI

RELATIONISM: AN ALTERNATIVE TO RELATIVISM?

In his prefatory notes to *Introducing Criticism at the Turn of the 21st Century*, Julian Wolfreys, the editor, defines the schools of critical theory emerging around 2000 in terms of "epistemological foci" (Wolfreys: 4). The turn to epistemology, to a form of Kantianism, signals the end of the critique of modernity, of the cult of rationality and the "grand narratives" it gave birth to in the Age of the Enlightenment. John Jervis, in *Transgressing the Modern* (1999), and Brian MacHale, in *Constructing Postmodernism* (1992), retrace the history of the departure from cognitivism and epistemology in the later nineteenth century and of its radicalization in modernism and postmodernism as the main chapter in the history of contemporary ideas.

The "ontological turn" pushed the rationalist critique of modernity's rationalism to its ultimate wreck on the shores of agnosticism. Previous forms of inquiry, such as "Is it possible for me to know the world?" or "How do I get to know this world?" yielded to that extreme form of relativism which is ontological uncertainty: "What world is this?"

The essays in Julian Wolfreys' introduction show the latest schools of critical thought endeavouring to rebuild epistemological reference frames, generally grounded in some form of relationism: the holistic picture of interrelated and emerging phenomena (chaos theory, complexity and other myths of the New Physics), ethical (oriented to the other) patterns of social life, diasporic consciousness, ecological modes of being in the world, holistic (hologrammatic), non-localized distribution of information, etc.

Is relationism something different or just a special case of relativism? Were agnosticism, latitudinarianism or "methodized" uncertainty (Heisenberg) corrupted forms of relativism, which has now newly resurfaced in a purged version as relationism?

VI.1. Looking backwards …

Relativism is taken to be the defining trait of modernity, as the end of fundamentalism, dogmatism and narrow-minded scholasticism. The pagan antiquity was revived, not in its letter, but in its spirit, as the Hellenistic free play of the mind around its object, which Matthew Arnold would see as the ancient prototype of modern objective criticism. The medieval way of appropriation—Virgil evangelized, for instance—had read its own ideology into the ancient source, invoked as the sine-qua-non "auctoritates" of any institutionalised discourse of the sermon and exemplum kind.

VI.2. Enabling Relativism

The Renaissance mind undertakes to restore the letter to its true spirit by setting the ancient picture in its own perspective. The new path Brunelleschi opened in the history of fine arts reclassified representation, not as Platonic *paradeigma*, a locus of the original's essential features, but as illusionistic game. It is *techne*, Plato's word for technical knowledge, skill or device that defines the new aesthetic of deception. With its reductive effect on visual experience, perspective drawing was not of the nature of *nomos* but rather of a hoax; it was a non-realistic mode of creating the reality effect.

Illustrative in this sense is Leonardo da Vinci's Vitruvian Man. The model set up by Vitruvius, the Roman architect of the 1st century BC, is not the archetype to be ritualistically repeated as IDEM, but a point of departure, the reference frame of repetition with a difference. The gap, and not the continuity, as in Plato, between reality and representation, creates a space of open-ended differentiation. The new art object emerges, not as autonomous or self-generated, but in relation to a precedent which it modifies. Its identity is conferred, not by repetition (observance of *auctoritates*) but by difference. Da Vinci differs from Vitruvius in the multiplication of perspectives. The male body is inscribed within two superimposed geometrical figures, a square and a circle, which creates possibilities for the geometrical increase of the space of representation (the combination of arm and leg positions creates sixteen different poses) in relation to the unicity of the space of being/ reality. The outstretched arms are no longer aligned with the naval but with the top of the head. The body of Prince Reason was centred on the head, as Juan Luis Vives fabled man (*Fabula de Homine,* 1518): the paragon of animals, possessed of reason and memory, looking before and after, rather than the hoarding, feeding and sleeping creature of bodily desire. Da Vinci's Vitruvian Man is not the locus of universal symmetries but rather a revelation of the relativism born simultaneously with the discovery of the laws of perspective.

The relaxation of the scholastic fibre was quickened in the sixteenth century by the new grammar school curriculum including books by Erasmus (*Adagia, De copia*), where comparative studies were encouraged (Livy's Lucretia in comparison to Ovid's etc).

The elitist and closed circles of initiation—the *en cyclos* format of the earlier *prisca philosophia* in Florence, or the Essex circle in England—revived what Vitruvius himself had used: the *analogia* or ancient Greek relativism in their treatment of esoteric lore. Myth and religion were deconstructed through amalgamation and codification. The mix of mythological traditions going back to the *prisci theologi* (Moses, Hermes, Zoroaster), to orphism, alchemy and the Kabbalah, emptied out theological notions, reducing them to signifiers sliding under signifiers in a semiological utopia of endlessly deferred cross referencing. It was such symbolic language that served Paracelsus or Nostradamus for the codification of their "prophecies."

VI.3. Lapserian Relativism

The late Renaissance or Baroque era felt the bitterness of blighted hopes. The century was closing without any of the hopes cherished by humanists being realised: no king had emulated the paragon set up by Erasmus, not even Archduke Philip the Fair to whom he had dedicated it. No commonwealth like the one imagined by Thomas More had acquired a geographical habitation and a name. The health of the body politic, which the rise of nation states and the religious reforms had seemed to foster, looked more like a fatal affliction in the midst of Counterreformation politics. Going back in time had a derealising effect. It was the deceptive force of device in art that came to the fore over life-like or realistic illusionism. Shakespeare's Edgar (*King Lear*) is as exuberant in the display of his deluding game on Gloucester as painter Girolamo Francesco Maria Mazzola (more commonly known as *Parmigianino*, "the little one from Parma") in the self-complacent demonstration of the distorting effects of his art in *Self-Portrait in a Convex Mirror*. Disappointment with the world forces the artist back on his own resources, both "artists" —Edgar as landscape painter through words and Parmigianino dressed in ermine and boasting the talent of his hand which looks bigger than the head—priding themselves on their skill of creating a parallel reality which

Edgar, at least, redeems by ascribing it a healing power in a world out of joint.

Back-stage politics had invalidated the Tudor myth sort of historical imagination, bridging Tacitism and Machiavellianism. Modern Fideism (Agrippa, Montaigne) was linking up with ancient Pyrrhonism. The IDEM, the repeat ideology, was pressing home a sense of the immanence of evil. Free play of differences, with its increase in potential symbolization, had deteriorated into conflicting narratives, which baffled the understanding. What was Hamlet to make of the superposition, not of collocating, non-conflictual representations of man, but of that chimeric body forced, on the one hand, into a constitutive frame provided by Pico della Mirandola's *Oration of the Dignity of Man* (1486) as "maker and moulder of himself," and, on the other, into a matrix or catalogue of human misery and afflictions listed in Pierre Boaistuau's *Le Théâtre du monde*? Life is a dream, society is a theatre of deceptive masks, and art is delusion. It no longer serves as a simulation model of the world; it is an isle full of noise or sirens' song that threatens to drive the intruder mad.

Medieval characterisation fixes individuals into social categories (estates). It is true that the Renaissance man is pulling out of his social location conceived of as statutory class in an attempt at "self-fashioning" (Stephen Greenblatt), but the quest for identity is still one that appeals to category, mediating concept or representation.

Shakespeare's "impressionistic" portraits are da Vincian superpositions of poses or constructions relative to the other characters in the play; confronted with forms of discursive ratification the moment he has given up on his statutory class, King Lear experiences the dizzying effect of his "incomposable" public self: Kent's "royal Lear" or "authority," Cordelia's royal father, Oswald's putative "my mistress's father," the clown's "Zero," his own devalued body whose hand "smells of mortality." Of psychological import is also what René Girard calls "triangulation of desire" (*Les feux de l'envie*). Unlike Don Quixote, who is cured of his attempt at self-fashioning, but whose aesthetic self is modelled on a textual (and outmoded) precedent, Shakespeare's characters are bearers of modern values, trying to unfix themselves from cultural stereotypes. It is the humanity of the ancient regime that is still sticking to degrees in schools, in city guilds, at court, etc. (*Troilus and Cressida).* The Queen represents Richard II as a second Adam; he sees himself as a providential vicar of God, etc. Does the individual face fit into the public visor? This visor is no longer given at birth. Shakespeare's sage (of a new, pragmatic school, though), Ulysses, claims that status needs public performance and ratification. Is sulking Achilles straying away from the battlefield? Then the soldiers will admire that "horse," Ajax, the unworthy hero of physical force. It is the relative positioning of characters that best defines them. The cast of the plays is usually caught in a transitional state that does not match the traditional hierarchy of statutory classes, while the new is not yet acknowledged. Prince Merchant Antonio is such a hybrid product of the rising economic classes within the bosom of an aristocratic society. He will despise Shylock who makes money from usury, that is, as unaccountable plus value, yet it is the same economic law and market mechanism that propel Antonio's cargoes along the dangerous lanes of the seas in search of profitable trade. As Salarino is ironically enquiring in the first scene, will these ships curtsey to one another as their master is probably still doing at the Duke's palace? Is not the Duke himself choosing an amiable and profitable relationship with Shylock over regard and compassion for aristocratic Antonio? Identity in Shakespeare is relative to perception by others, public performance and construction, and, therefore, the most conspicuous shifting ground at the onset of modernity.

VI.4. The Birth of the Semiotic Consciousness.

Relativism or relationism has a destabilizing effect on identities. The law of gravity was merely the force that prevented the atoms rolling through the void from pulling apart and disintegrating the sea of matter. It only acted from without, leaving unaffected the constitution of the entities existing in the universe. Newton's materialistic, mechanicist, atomistic theory reinforced the reassuring substantivalism of previous centuries.

The enemies of mechanical empiricism in the early modernity, who found it reductive and vulgar, searched for proofs, not of an oppositional metaphysics, but of some form of relationism. Forced into a coded bond with the Other, the individual entity acquired only some participatory, putative or differential form of identity.

In the beginning was Shakespeare, whose definition of identity in *Troilus and Cressida* is less a matter of the deconstruction of medieval statutory classes into simulacra (see Jean Baudrillard, *Symbolic Exchange and Death)* than a classification of social roles through coded differences (degree):

> Degree being vizarded.
> The unworthiest shows as fairly in the mask.
> The heavens themselves, the planets and this center,
> Observe degree, priority and place,
> Insisture, course, proportion, season, form.
> Office and custom, in all line of order: (I/3)

The same social category—say schoolmasters or merchants—has inner divisions according to degree, and it is not persons but bits of social meaning or types of public performance that are coded. It is fighting on the battlefield that earns applause, not the empirical self: once out of sight, heroic Achilles will be replaced in the soldiers' admiring gaze by brutish Ajax. Degree, priority, place, insisture, course, proportion, season, form, office and custom are social ciphers, not natural differences. Lear is a fool for substituting quantifiers and natural language for the contractual "bond," an element of the coded father/daughter relationship Cordelia invokes in reply to his questioning of his daughters' affection for him. Bonding is a matter of language, the breaking of the vows and thereby of her bond to Troilus evacuating Cressida from syntax: Broken or incomplete syntactic patterns, such as "Cressida is … does …" convey Troilus's sense of Cressida's loss of defining attributes (predication) and role (appropriate action).

Edgar will teach Edmund "differences," those that matter in the constitution of the category of nobility, and he does not mean by that the material issue of the estate the bastard is craving. Shakespeare's humanity is Luhmann's stratificational society, which builds on nature layers of coded meanings (*Die Gesellschaft der Gesellschaft*).

Whereas Shakespeare welcomed the emergence of a symbolic instead of substantival order, John Donne read the stand-in as empty prop, as lacking being. The new social dignity of ambassador in the dawning age of diplomacy impressed him as derealization of one's self as signifier. Unlike Henry Wotton's diplomatic mission, Donne felt that his poetry granted his autopsychological public performance discernibility, or rather he liked to think so, for an undercurrent of scepticism engulfed his very confidence in the possibility of identitarian narratives ("If there be such a thing as I") at a time when referential identity was no longer full but a language game turning up signifiers for other signifiers. The king's own identity was rendered metonymic by the relation of equivalence between himself and the ambassador, sealed by the credential letter:

TO SIR HENRY WOTTON AT HIS GOING AMBASSADOR TO VENICE.

AFTER those reverend papers, whose soul is
   Our good and great king's loved hand and fear'd name;
By which to you he derives much of his,
   And, how he may, makes you almost the same,

A taper of his torch, a copy writ
From his original, and a fair beam
Of the same warm and dazzling sun, though it
Must in another sphere his virtue stream;
....

Admit this honest paper, and allow
It such an audience as yourself would ask;
What you must say at Venice, this means now,
And hath for nature, what you have for task.
....

But 'tis an easier load, though both oppress,
To want, than govern greatness, for we are
In that, our own and only business,
In this, we must for others' vices care.

'Tis therefore well your spirits now are placed
In their last furnace, in activity;
Which fits them—schools and courts and wars o'erpast—
To touch and test in any best degree.

For me—if there be such a thing as I—
Fortune—if there be such a thing as she—
Spies that I bear so well her tyranny,
That she thinks nothing else so fit for me.

The relationship between England's king and his stand-in abroad is charged with ambiguities that are mutually destabilising, as suggested by the thick language and loose syntactic structure: "… By which to you he derives much of his …." Donne is spinning metaphors for the surrogate identity of the ambassador and its incompleteness ("makes you almost the same"): not even as a copy, a double, is his status a satisfactory one. An ambassador is but a metonymic or proxy version of the original: "a taper of his torch," "a copy writ of his original." "His original" signifies loss of origin, actually. The king's personhood has split into origin and derivative selfhood, which bears the trace of the original's adulteration caused by the king's deferral of part of it to an authorised form of his otherness: the ambassador's mandated identity.

By contrast, loss of the weighty clod of referent seems to have impressed Shakespeare as the very premise of creating a brave new world, not through material processes, such as the commoners' uprising or the bloody deposition of kings, but through the introduction of a new symbolic order for the assessment of human worth and the good life. The

Ulysses of *Troilus and Cressida*, the traditional figure of wisdom, knows that reality will always be *à sa place*, but also that a language that is no longer tied down to the pre-existing reality of estate, birth and rank can write/speak/perform another into being. The medicinal show of the Greek commanders' feigned disregard for Achilles which he is staging, as well as his own sermon passed off as a letter from "a strange fellow" are counterfactuals, yet they manage to bring Achilles back to the battlefield. As well as Edgar's medicinal hoax on his father or Ophelia's feigned reading, language acts no longer possess either metaphysical or ontic relevance.

In his attempt to understand the Greek leaders' changing attitude to him, Achilles is falling back on a narrative: the medieval view enshrined in the *De cassibus* tragedy. Fortune may dictate changes in one's material status, such as the loss of "place, riches, favour," which are but accidental signs of prestige ("standers"). They are just props of a surrogate concept of worth whose true ground ought to be talent. Identity is a social construction, not a form of embodiment: one does not accrue honours to himself simply by being man, the paragon of creation. His status in society is a matter of external ("without him") ascriptions, yet they are reified, realised as a "system of objects" (Baudrillard). Achilles is speaking on behalf of the humanists of yesterday when he places talent above the feudal marks of statutory classes (birth, rank, estate), while also clinging to a theological view of earthly life being controlled by some metaphysical agency.

> ACHILLES
> What, am I poor of late?
> 'Tis certain, greatness, once fall'n out with fortune,
> Must fall out with men too: what the declined is
> He shall as soon read in the eyes of others
> As feel in his own fall; for men, like butterflies,
> Show not their mealy wings but to the summer,
> And not a man, for being simply man,
> Hath any honour, but honour for those honours
> That are without him, as place, riches, favour,
> Prizes of accident as oft as merit:
> Which when they fall, as being slippery standers,
> The love that lean'd on them as slippery too,
> Do one pluck down another and together
> Die in the fall.
> —*Troilus and Cressida* III/3

What is really sliding, according to Ulysses who is walking in "reading," is the signifier under signifiers in communicational contexts. Achilles believes in the unproblematic reflection of a stable form of identity in the other's eye, that bodily presence will guarantee the nature of the spirit inhabiting it:

> The beauty that is borne here in the face
> The bearer knows not, but commends itself
> To others' eyes; nor doth the eye itself,
> That most pure spirit of sense, behold itself,
> Not going from itself; but eye to eye opposed
> Salutes each other with each other's form;
> For speculation turns not to itself,
> Till it hath travell'd and is mirror'd there
> Where it may see itself.
> —*Troilus and Cressida* III/3

Ulysses is sceptical of stereotypes: Achilles has just pronounced a commonplace, the position is "familiar" to him, but truth is circumstantial and relative to the illocutionary source. The writer of the letter Ulysses pretends to be reading, blaming on him his own unorthodoxy, relativizes status, making it dependent on "circumstance." Identity is not essence, but it is not rooted in a general, stable and objectified institutional model either. It is circumstantial and emergent in dialogue, as Hans-Georg Gadamer would later define meaning in general. A communicational model which defines the individual in relation to an otherness in terms of performance and circumstantial assessment by local interpretive communities replaces both the humanists' self-fashioning or the prescriptive, stereotyped portrait of conduct books. The material or spiritual "consisting," the full presence, vanishes into a metonymic presence of "parts," a spurious reality effect (heating of steel) and the others' signs of recognition: "applause." The individual has no control over his public perception (*no man is the lord of anything*); in fact he is *noman* until he is perceived, Shakespeare's Ulysses claims, erasing the universalist assumptions of the medieval "Everyman" and anticipating Berkeley's extreme idealism. The applause is not the confirmation of physical reflection but a forming, shaping agent. "Can anybody tell me who I am?," Lear inquires, aware of the identitarian effects of the others' definitions of his person and role. Identity is becoming a mobile, mutating reality. The Ajax signifier is sliding from the "most abject" and "lubber" under the "renown'd" signified.

Be that as it may, and airy and insubstantial as they are, the semiotic scripts of art, based on honest deception, as they are excused by a good end, can perform a healing function which shows social norms to be comparatively inefficient or oppressive. Montaigne's relativism had taken him all the way to Fideism—read agnosticism. There was one truth on one side of the Alps, and another on the other (*The Apology of Raymond Sebond*). No Augustinian churchman had gone farther in the nullification of life's worth. As nothing was new under the sun, to live for just one day would be enough to get the whole of the world's picture. No need for everyman to endure the blows of an invariably wretched life (Montaigne, *For to Philosophise is to Die*).

Shakespeare's Hamlet, as a Wittenbergian student, will inquire into the truth of narratives about the good life, which, at the time when Shakespeare was writing the play, stood in an uneasy relation to one another. The Pico della Mirandola tradition of the greatness of man had recently been reinforced by Juan Luis Vives's *Fabula de Homine*, but so had been the dirge on man's wretchedness breeding contempt of the world, which Pope Innocentius III published at the close of the twelfth century by John Gascoigne's recent English translation (1576). The two counter narratives representing the human being for the late Renaissance man are invoked by Denmark's prince early in the play as a superposition of opposite states and therefore one of the reasons for his disturbed state of mind. Contemporary Pyrrhonists and Fideists did not seem to be troubled by mind's incapacity to say that something is either true or false, but Hamlet, after an initial and wrong decision to erase from his mind the contents fed into it by his youthful empirical experience of the world in favour of received ideas or ghostly texts copied onto the tablets of his mind, undertakes to prove the worth of words in Baconian fashion. He frees himself from idols, experimenting with heuristic dramatic representations, including his self-dramatization, and finally manages to sort out the handful of dust version of man from its "discourse of reason" counter, which makes him the paragon of creation. It is the body, even the body of mundane glory, like Conqueror Alexander's or Emperor Caesar's, that comes down to the handful of dust which fills a hole in a barrel. The dust is a sign of what it displaces, i.e. of emptiness, apart from being the material Adam was made of, according to the Biblical script reproduced by Erasmus in his Enchiridion. On the contrary, the sheep or goat skin, serving as parchment for manuscripts, seems to fare better. Writing is a weaving of signs which carry the discourse of reason through time; it is the wine in the barrel, the wine that is Christ's blood, or the resurrecting spirit out of the dying body. Hamlet will be redeemed by Horatio's story. It will do justice to a Hamlet of immaculate conception, born of the word, written into being against the world's rest of silence, like a new cosmic order emerging out of chaos. The "hero" signified is sliding from the Alexander and Caesar under the Hamlet signifier.

The passage from medieval dogmatism to modern pluralism bred a dangerous form of relativism verging on agnosticism and suspended judgement. Shakespeare solves the dilemma of contradictory truths or values by inserting the fable of man into relational frames. For instance, man is impotent in relation to his mortality, but great on account of his infinite faculties and unique condition as a meaning being. The emerging society of print, letters, images, ciphers, commodities, plus value, etc. was a semiotic web of infinite interconnectedness.

The Shakespearean man is embedded, not in a universal but in a social and historicized frame, his condition being conjugated with the whole of social semiosis. He understood society as Niklas Luhmann would later write about it in *The Society of Society (Die Gesellschaft der Gesellschaft*). Man's world is superimposed on nature as sets of systematic differences from it.

> That is why autopoiesis is not to be understood as the production of a specific structure [Gestalt]. What is decisive is much more the creation of a difference between system and environment (Luhmann: 66)

Individual consciousness does not depend on another—"salutes each other with each other's form," as Achilles naively believes—but on a "higher order system, that does not itself live," which is "the communication system of society," the symbolic order which fuses culture and society. Ulysses is not interested in the nature of bodies, for Ajax may very well be a horse, yet, by fighting as the Greeks expect a hero to do, he gets universal acclaim. Perception draws distinctions between the two warriors, positioning them according to "degree."

Ulysses reading his own text is less the embodiment of sage discourse —in fact, what he says sounds neither wise nor lofty, but pragmatic, almost cynical—than what Luhmann would call society's "autological" text that integrates distinctions made in the social, temporal and functional dimensions (880). Ulysses tells the Greek leaders of the armies what their society chose to function with, what network of distinctions was being successfully communicated across its constitutive groups.

A current criticism of Shakespeare's drama is that his characters are a jumble of conflicting discourses from which the playwright remains serenely distant, entrusting the audience maybe with the task of assessing their relevance.

It is true, as Nicholas Marsh remarks (Marsh: 16), that the dialogue keeps jumping from one idea to another in an unexpected fashion, that hypothesis are launched which lead to no conclusion, but the effect on the audience is the calculated impression that we have been plugged into the heteroglossia of everyday communication whose effect is differentiation, the emergence of coded, indexical conduct and decreed identities (visors). Both Ophelia and Achilles fail to notice that "at the higher levels the importance of the individual diminishes." Ophelia still isolates Hamlet from all his subjects, raising him to a unique position of social eminence, where he becomes the "observed of all observers," the model emulated by the whole people. Achilles does not realise the alienating effect of the gaze of the other. It is not an impassable mirror but a shaping agent. In the others' eyes, Achilles can only see himself as represented, and this representation is not an essence or set of features, but an arbitrary public self. We have just witnessed Achilles' misreading of his peers' feigned slighting of his person. He is reading what Ulysses had put there, as Ulysses himself is reading his own ideas into the presumably received letter.

> For practical purposes . . . events occurring in other persons are accessible to an observer in terms of inference alone; all he observes is the stimuli which reach the other person and the latter's reactions; the rest is subject to conjecture. Furthermore, the observer, being a social stimulus for others, possesses knowledge about the origin and the nature of some of the stimuli which he feeds to other individuals. In such a system, which includes the observer as an integral part, the actions of the first person are stimuli for the second person and the response of the second person are stimuli for the first person (Ruesch & Bateson, 26)

From his higher level of perceptibility, similar to Dupin's in Lacan's reading of "The Purloined Letter," Ulysses accesses intersubjectivity, the space where he gets from the other his own message in reverse form. His deceptive scheme works perfectly, Achilles being brought round to the right (society's) point of view.

According to Luhmann, this set of distinctions whereby society defines itself is the highest form of consciousness. Shakespeare sensed the very mechanism of society's autopoiesis. It was not the individual's self-fashioning but the very heart of humanity's mode of existence that kept his mind busy until the secret was opened.

VI.5. Relationism from Leibniz to Quantum Darwinism

The Newton/Leibniz divide over the absolute/relative and supervenient nature of space and time has been radicalized in our time by the heated arguments supporting, on the one hand, the existence of a timeless universe, and, on the other, the evacuation of objects which disintegrate into processes from a space temporalized as flux. At one end of the spectrum, Julian Barbour sees instances of time as disconnected photographs, static and timeless; at the other, David Peat proclaims the end of representation and the onset of uncertainty, incompleteness and chaos.

Without going into such extremes, Carlo Rovelli (Rovelli 1637) and W. H. Zurek (Zurek: 181-188) come up with relational and contextual descriptions of quantum mechanics which are proved to work much better than the context-invariant versions. Rovelli agrees, therefore, with Zurek that "properties of quantum systems have no absolute meaning," and that they must therefore be always characterised with respect to other physical systems. Zurek himself thinks that "the correlations between the properties of quantum systems are more basic than the properties themselves," which should make sense in the context of the Einstein-Podolsky-Rosen experiment of 1935, which proved the existence of such correlations between quantum systems from a distance.

Philip Anderson's revolutionary contribution to condensed matter physics was that of invalidating the myth of Eigenstates as stable features of systems. Changes at the microscopic level, due to impurity-induced processes of the particles' localization, will lead to changes at the macro level, including the Eigenstates. The definition of identity through context-free invariants has been regarded with suspicion since Anderson's publication of his "More is Different" article in 1972, while the alternative view, that everything in the universe is more than itself, entering into some form of relationship with the environment, has been gaining ground.

According to Carlo Rovelli, in quantum mechanics "states" as well as "value of a variable" or "outcome of a measurement" are relational notions in the same sense in which velocity is relational in classical mechanics. Thus, a quantum mechanical description of a certain system (states and/or values of physical quantities) cannot be taken as an "absolute" (observer independent) description of reality, but rather as a formalization, or codification of a system's properties relative to a given observer.

Moreover, physicists of various commitments are now agreed that "the classical world arises from an ultimately quantum substrate" (Zwolak 2009: 110402-1). The process consists of exchanges of information between system and environment, which leads to decoherence of the system or collapse of the wave function (the systems collapses to a pointer state or Eigenstate).

We may compare the quantum Darwinists' "plateau" (Zwolak 2009: 110402-3) to Philip Anderson's localization and Deleuze's "territorialisation" or regime of signification theorised in *A Thousand Plateaus,* to be able to say that a phase of culture has reached maturity and stability, being characterised in similar ways by all observers. That is the moment, according to Zwolak, Quan and Zurek (Zwolak: 2009: 110402-1), when the leak of the system's information into the environment reaches a point where the transfer remains constant. We are likely to get the same information from almost any position in the environment.

Codification is also the main feature of a quantum system's opening to the world, its change depending on exchange of information with the environment, which organizes itself into a system on the basis of the "intercepted" information, while the system is dismantled from informational plateau to the environment of the new, emerging system. Anderson calls this process delocalization; Deleuze, deterritorialisation.

Deleuze may have been influenced by Anderson's theory, but his acknowledged affiliations include Leibniz, whose views of relativity and relationism are perhaps slightly distorted to suit the French philosopher's own speculations around repetition and difference:

> This is what the term "vanishing difference" means: it is when the relation continues even when the terms of the relation have disappeared. Now it is here that Deleuze reaches, and attempts to formulate, a principle of pure difference, at least insofar as it is expressed Leibniz's formulations of the differential calculus. The differential relation can be said to be a *pure relation*, insofar as it is a relation that persists even when its terms disappear: it thus provides him with an example of what he calls the concept of *difference-in-itself.* Normally, we think of difference as a relation between two things that have a prior identity ("x is different from y"). With the notion of the differential relation, Deleuze takes the concept of difference to a properly transcendental level: the differential relation is not only *external* to its terms (which was Bertrand Russell's empiricist dictum), but it also *determines* its terms. In other words, difference here becomes *constitutive* of identity, that is, it becomes productive and genetic. This is what Deleuze means, in *Difference and Repetition,* when he says that relations such as identity, analogy, opposition, and resemblance are all secondary *effects* or results of prior relations of difference (Smith:16-17)..

Leibniz, therefore, defines difference as Derridean *trace*: the very site of a type of relation in which a process—of differentiation—takes priority over the things involved in it. These relations reach out in all directions, and that is why all the individual monads may be said to be embedded in the all-comprising, divine monad. Like the uniform distribution of information in a hologram, whose whole image appears in any splinter in case the plate gets broken, the principle of sufficient reason allows any subject to stretch out backward and forward until it includes the whole world. Daniel W Smith finds Leibniz's concept of expression "fantastic," yet the philosopher is thereby defining a semiotic utopia, in which a sign refers to all the others.

If we think of the three-dimensional image of the bidimensional plate of a hologram, we may say that there is nothing "fantastic" about the non-localization of the principle of inclusion in Leibniz's infinite analysis of the domain of existences either:

> What is the difference between truths of essence (principle of identity) and truths of existence (principles of sufficient reason and indiscernibility)? With truths of essence, says Leibniz, the analysis is finite, such that inclusion of the predicate in the subject can be demonstrated by a finite series of determinate operations … The analysis of truths of existence, by contrast, is necessarily infinite: the domain of existences is the domain of infinite analysis. Why is this the case? Because if the predicate "sinner" is contained in the concept of Adam, then if we follow the causes back and track down the effects, the entire world must be contained in the notion of Adam. When I perform the analysis, I pass from Adam the sinner to Eve the temptress, and from Eve the temptress to the evil serpent, and from the evil serpent to the forbidden fruit, and so on …. (Smith: 16).

Glossing on the concepts of difference and repetition in Deleuze in contrast to Aristotle's specification of sufficient reason, Daniel W. Smith identifies the common link between the latter and Leibniz's theory of discernibility:

> Leibniz is able to reconcile the concept and the individual only because he gives the identity of the concept an infinite comprehension: every individual substance, or monad, envelops the infinity of predicates that constitutes the state of the world. Where the extension of the concept = 1, the comprehension of the concept = ∞ (an actual infinity). It is one and the same thing to say that the concept goes to infinity (sufficient reason) and that the concept is individual (indiscernibility). In pushing the concept to the level of the individual, however, Leibniz simply rendered representation (or the concept) infinite, while still maintaining the subordination of difference to the principle of identity in the concept. (Smith: 12.)

Much has been achieved closer to our time by finite representations of the infinite, such as the straight line "bended" into a circle (Fig. 13) or the Riemann sphere (a two-sphere whose points are identified with all complex numbers by a stereographic projection), where the possible representations of a thing may reach the infinity symbol at the top (Fig. 14).

Contemporary studies of non-linearity are analogous to the differential calculus whereby Leibniz was trying back then to relate the apparently incompatible orders of symbols and reality in a way which allowed him to discover the deep structure of reality (the existence of a deep mathematical structure in a stratificational universe has become a commonly accepted idea) through the vanishing yet constitutive differences.

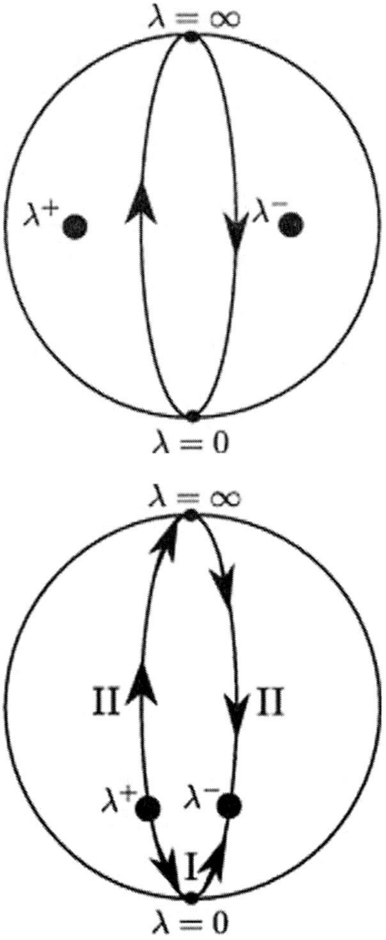

Fig. 13.

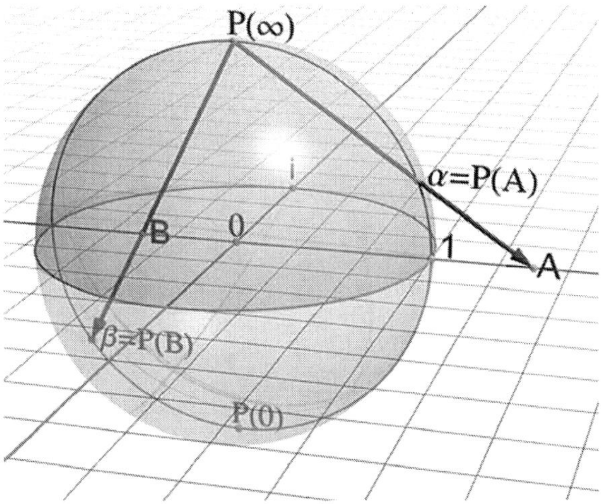

Fig. 14

Unlike Newton's notions of absolute space and time, Leibniz conceived of them as orders of things (simultaneous and successive), which could be computed as they obey laws of possible assemblage ("possibilia") and transformation, which is as much to say as that a system's evolution depends on initial conditions, and that time and space are interdependent, a later state being supervenient on a previous one. Between the spatial order of t2, for instance, and that of t1, the change, although not predictable in proportion of one to one, can however be probabilistically circumscribed through what we call (Lorenz's) rules of transformation.

The letters exchanged by Newton and Leibniz, published in an 1850 edition of *Leibnizens Mathematische Schriften* (edited by C. I. Gerhardt) show the former writing down dizzying rows of equations and figures. Unlike Newton, whose attention was divided between close observation of empirical phenomena and pure abstractions, Leibniz seems to have been pursuing through differential calculus a purpose which comes closer to semiotics than studies of non-linearity. Writing from Hannover to the more affined Hugens in September 1679, Leibniz confesses his dissatisfaction with the microscope which, although capable of revealing the hundred thousandth part of an object, is not of much use "pour penetrer dans l'interieur des mixtes" (Leibniz a: 22), i.e. of really existing things.

The mathematical analysis he proposes, which will cut across geometry and algebra, has the effect of evacuating the material bodies in order to reveal a non-localized order of symbols, a holographic distribution of information which conceptualizes the individual while also making him a site of the whole (superposition of individual and divine monad). In a homogeneous space, "tous les points du monde ont de la congruité entre eux, c'est à dire l'un se peut tousjours mettre à la place de l'autre." (a: 23). Y is congruous with (Y). Leibniz gives examples of homologies which are not affected by the size of the movable triangle or segment. The relation is preserved even when the related elements vanish (for instance, when the tangent to the sharp angles of an orthogonal triangle will at some moment reduce the orthogonal angle formed by Δ y crossing Δ x to the 0 of the intersecting axes).

Leibniz proceeds by studying infinite interconnectedness and minimalist reduction to the individual who, by being set in perspective and relations, becomes congruous with the whole world which is projected in himself (the geometry of congruity expanding from point to line, surface, sphere).

Leibniz was the first world builder, followed by d'Alembert's similar attempt to combine disciplines (geometry and algebra) in order to pass from Aristotelian logic to probabilistic logic which applies not only to what there is but also to what might have been and under what constraints could other states of the system pass from virtual/possible to real. The best of all possible worlds is that in which discords and dissonances are resolved into ultimate harmony. The mix of chance and determinism, or the rationale of complexity (self-organizing systems) are now current descriptions of the universe, or rather of what Deleuze derided as "chaosmos."

Conclusions

The New Physics represents the universe as emergent and consisting of non-linear, interrelated processes. Essentialist and invariantist positions (analogous to the principle of essence versus existence in Leibniz) have drastically diminished in number in favour of contextual and communicational models. Relativism is still surrounded by an aura of anxiety, especially in light of two world catastrophes of the last century, which revived an intense interest in ethics, but relationism, whose name was coined by Karl Mannheim, is the epistemological paradigm of the age of computers and of semiotic aesthetics.

It has been customary to define modernism and postmodernism as the transgressing of modernity (John Jervis), and the critique of modernity's grand narratives. As Hélène Védrine argues in her *Philosophies de la Renaissance*, there was a dark as well as a light side of the Renaissance. The enthusiasm stirred by the new discoveries and technologies and the rise of empiricist philosophy was counterbalanced by an esoteric school of the occult centred on Italy. Its purpose, however, was no longer mystic or religious. On the contrary, the mix of traditions had a demythologising effect, the underwriting agenda being a sort of *mathesis universalis*. The combination of elements from the Hermetic doctrines, alchemy, gnosis, or the Kabbalah revealed analogies which suggested the similar workings of the archetypal imagination rather than mystic revelation. The interest which Leibniz himself took in Renaissance Neo-Platonism was probably motivated by his fascination with that congruence which was the source of a holographic consciousness. Similarly, analogy, which he was exploring in geometry through moving lines and figures, could also be seen at work in the homologies of alchemist, Cabalist, gnostic-Biblical and alchemical permutations of the myth of the fall of the spirit into matter and its resurrection, for instance. As in a hologram, the information of the whole is contained in the part; the divine into the individual human monad. Leibniz's semiotic, infinite deferral of signifiers brings him close to our age of cybernetics, i.e. of "interconnectedness" and hypertextuality.

Védrine's interlaced narrative of official and alternative cultures also applies to the movement of ideas in the following two centuries. The Descartes-Newton-Locke-Voltaire filiation of the rationalist, absolutist and empiricist frame of mind was only the official and canonical surface of modernity. The dark side of relativistic thought and of speculations in the field of probability is now building an arch over time, connecting with the twentieth-century physics of nonlinearity, self-similarity, self-organization, interaction and emergence. The "autopoiesis" of modernism and postmodernism spiralled away from the modernity of the Enlightenment, while bringing the germs of its relativistic underside to fruition. The Leibniz revival is proof of a growing awareness of filiation among contemporary critical theorists, philosophers, and scientists of the probabilistic school for whom only the unseen, hatched by the mind, can guarantee the seen.

The solipsism of the mind from Immanuel Kant's categorical imperative to the unreturned gaze of modernist observers intoxicated with the vertigo effect of regressive self-contemplation (Valery's *Je me voyais me voir)* has receded into the history of ideas, post-war anthropology forcing the ecological mind into the spotlight:

> 'The individual mind is immanent but not only in the body. It is immanent also in the pathways and messages outside of the body; and there is a larger Mind of which the individual mind is only a subsystem ... immanent in the total interconnected social system and planetary ecology. (Bateson a: 467).

The notion of "perception of perception" only makes sense now as "silent agreement," "mutual influence" and "mutual recognition." The "irreducible face of the other" (Emmanuel Lévinas) is the source of the self's ontology in dialogue.

Bateson even speaks about an ecological aesthetics whose task is the identification of patterns, writing itself having become a sort of probabilistic calculus:

> Causal explanation is usually positive. We say that billiard ball B moved in such and such a direction because billiard ball A hit it at such and such an angle. In contrast to this, cybernetic explanation is always negative. We consider what alternative possibilities could conceivably have occurred and then ask why many of the alternatives were not followed, so that the particular event was one of those few which could, in fact, occur. (Bateson a: 405)

The global net of interfaced computers is *mathesis universalis* in an artificial language such as Leibniz, disappointed at the imprecisions of the natural ones, was planning to invent, as he confesses in a letter to Hugens. Taken up by Vannevar Bush around the mid-twentieth century, the project has finally been realised, its rhizomatic, lateral linking being known as "hypertext." However, the process that makes the whole scheme work might well be called "vanishing difference," by which Leibniz meant the infinite relationism reducing the order of things to an intelligible structure, whose centre is everywhere and the circumference is nowhere.

References

Bateson, Gregory. (a).1972. Steps to an Ecology of Mind: Collected Essays in Anthropology, Psychiatry, Evolution, and Epistemology. San Francisco: Chandler Publishing

—. (b). 2000. 'Form, Substance and Difference' *in Steps to an Ecology of Mind* .Chicago: University of Chicago Press

Leibniz, G. W. (a) 1850. *Leibnizens Matematische Schriften,* ed. C. I. Gerhardt Berlin: Verlag von A. Asher & Comp.

Leibniz, Gottfried Wilhelm. (b) 1687. Letter to Antoine Arnauld, 19 October 1687. *Leibniz's Correspondence with Antoine Arnauld*. Early Modern Texts - Gottfried Wilhelm Leibniz Philosopher http://www.earlymoderntexts.com/pdfbits/leiba3.pdf

Luhmann, Niklas. 1997. *Die Gesellschaft der Gesellschaft*. Frankfurt am Main: Suhrkamp.

Marsh, Nicholas. 2003. *Shakespeare: Three Problem Plays*. New York: Palgrave Macmillan.

Peat, David. 2002. *From Certainty to Uncertainty*. Joseph Henry Press.

Rovelli, Carlo. 1996. "Relational Quantum Mechanics." *International Journal of Theoretical Physics*. 35: 1637.

Ruesch, J. & G. Bateson. 1968 /1951. *Communication: The Social Matrix of Psychiatry*. New York: W. W. Norton & Company

Smith, Daniel W. *Deleuze on Leibniz: Difference, Continuity, and the Calculus* http://m.friendfeed-media.com/9f13c7b42c059ecc66c4b22d363f4d9c6ab61de5

Wolfreys, Julian (Ed). 2002. *Introducing Criticism at the 21st Century*. Edinburgh: Edinburgh University Press.

Zwolak, Michael, H. T. Quan, and Wojciech H. Zurek. 2009. "Quantum Darwinism in a Mixed Environment." *Physical Review Letters*. 11 September 2009. PRL 103, 110402-1 - 110402-4.

INDEX